PRAISE FOR
From Here, You Can't See Paris

"[Sanders] renders the restaurant's workday as cannily as he does the village's moments of abrupt dislocation from the present, when the air suddenly seems to hold a thousand years of history in it. A good and leathery year abroad, an honest and deeply enjoyed experience that avoids skimming off only the fruity bonbons while neglecting the ruck of daily life."
—*Kirkus Reviews*

"*From Here, You Can't See Paris* is a wonderful book that captures rural France."
— *Portland Oregonian*

"[A] mouthwatering story."
—*Dallas Morning News*

"Sanders' account is so lovely, and Les Arques so sensuous and ripe with magic."
— *Publishers Weekly*

"A delightful, sympathetic, and revealing journey into the heart and stomach of the new provincial France. More enjoyable and instructive than a recipe book or a travel guide."
—THEODORE ZELDIN, author of
The French

"*From Here, You Can't See Paris* is a delightful tale of how an American discovers *le pays*, the little France where the roads on the Michelin map are the thinnest of lines. It is the story of a young couple's restaurant and the ancient French village that it brings back to life. Rather than idealize what he sees, Michael Sanders' report is all the more memorable for its clear-eyed honesty."
— PATRIC KUH, author of
The Last Days of Haute Cuisine

"Michael Sanders provides a truly intimate, in-depth, and eloquent portrait of rural life in a beautiful part of southern France. Rather than the familiar rosy sentimentalism that characterizes many other books in this genre, his caring and warm admiration for the villagers of the Masse and Lot valleys and his understanding of their arduous lives and time-honed skills adds a refreshing dimension of authenticity to this intriguing and evocative work."
— DAVID YEADON, author of
Backroad Journeys of Southern Europe

"Michael Sanders, in his extraordinary tale, has weaved a complex, rich textural tapestry of everyday life in the Lot that is honest, funny, and endearing. I have a summer home there and it all rings true to me. He writes with warmth and affection as well as precise, keen observation. This is a book for anyone who loves France or who wants to travel there."
— KEN HOM, BBC-TV presenter and author

About the Author

MICHAEL S. SANDERS, a former book editor
and author of *The Yard*, lives in midcoast
Maine with his wife and daughter.

FROM HERE, YOU CAN'T SEE PARIS

ALSO BY MICHAEL S. SANDERS

The Yard: Building a Destroyer at the Bath Iron Works

FROM HERE, YOU CAN'T SEE PARIS

SEASONS OF A FRENCH VILLAGE AND ITS RESTAURANT

MICHAEL S. SANDERS

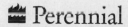 Perennial

An Imprint of HarperCollins*Publishers*

First Perennial edition published 2003.

Designed by Christine Weathersbee

Map illustration by Jackie Aher

The Library of Congress has catalogued the hardcover edition as follows:
Sanders, Michael S.
    From here, you can't see Paris: seasons of a French village and its restaurant / Michael S. Sanders.
        p. cm.
    Includes bibliographical references.
    ISBN 0-06-018472-8
        1. La Récréation (Restaurant). 2. Food habits—France—Les Arques. 3. Les Arques (France)—Social life and customs. I. Title.
TX945.L35 S26 2002
394.1'0944'733—dc21   2002023288

ISBN 0-06-095920-7 (pbk.)

    05 06 07 ❖/RRD 10 9 8 7 6 5

Pour Jacques et Noëlle Ratier

CONTENTS

ACKNOWLEDGMENTS

I would like to thank all the people of Les Arques, and those in the surrounding villages of Cazals, Moncléra, Marminiac, Gindou, and Goujounac and elsewhere in the Lot who helped us on our quest to adapt and prosper.

Jacques and Noëlle Ratier are, and now always will be, generous, spirited, lovely friends, the only kind worth having. There is no mystery why their restaurant is such a success; it is no more than a reflection of two very good people.

Jean and Anne-Marie Winter, indirectly through Marion Knox, found La Récréation for me, and never has a restaurant recommendation mattered so much in the life of one person!

Raymond and Suzanne Laval; Alain and Michelle Laval and their children, Yann and Melvin; and Gérard Laval—this was a family always there for us, and who opened their lives and homes to us in many hours of need. Our life would have been lacking in so many things without them.

Up in the village, I would have never gotten to know Les Arques and its past without the help of Elise Ségol, Blanche Delfort and her daughter Simone Bladié, Sylvie Lacombe, Guy Fillion, Patrick Cantagrel, and Lionel Gramon.

At the restaurant, Jacques' mother, Jeannette, a big-hearted jokester, was just one in a crowd of able help always ready to answer a question, always saving a tidbit for me to savor, always happy to explain what and why to someone who was, after all, a stranger. Bénoit David, Frédéric Dudon, and Sandrine Gillet were particularly helpful. Jeanette Branwhite, capable, knowledgeable, endlessly hospitable, and with a word or fourteen in answer to any question, eased much of the pain of adjusting and opened her home and her circle to us when we hungered for just an evening of English. Tim, Daniel, and Nats completed that warm circle.

Guy and Martine Cousin supplied wry commentary and filled in the blanks, political and otherwise, others did not or could not know. Yvonne and Marie-Josée Trégoux, Christiane Longi, and the Dézys are the kind of neighbors you dream about having when you move to a foreign land. Bernard Bousquet, Luc and Ghislaine Borie, Alexis Pélissou, Gilles and Laurent Marre, Geneviève Crassat and Pierre-Jérôme Atger, Françoise and J. Albert Reix, Madame Delon, La Famille Jouffreau, Jean-Luc Baldès, Philippe Burnède, Pierre Sourzat, Jean-Marc Dubiau, Albert Sanchez, Thierry and Luze Deschamps, Ken Hom, Jacques and Pierre-Jean and Babé Pébeyre, all experts in areas from truffles to goat cheese, gave generously of their time and of their selves to help me to understand the mysteries of their worlds.

The Loftie and Forshall families, who have long lived and worked in this part of France, provided a different perspective, as well as more kids and dogs to entertain us all. To Chuck and Alice and James and Kathy, we lift a glass of Cahors to sunsets in Collioure and children's birthday parties, and may we enjoy both together again.

For taking Lily in hand, ushering her painlessly and with true heart into the world of French, I thank Sandrine and Isabelle at the Café de Paris, Claude and Lydia and Noémi of the Epicerie Bouysset, Madame Stuker at the market, Monsieur and Madame Miguel at Casino, the Alazards at the *tabac*, Madame Crochard at Bruno's bakery, the staff of the École Maternelle at Marminiac, and all those good souls who encouraged her efforts and cheered every step forward.

Finally, credit where credit is due, the man who first inspired me to speak another language, and to speak it well, to explore the mysteries of the literature, food, and culture of France, was Denis Asselin, my French teacher at Westtown School.

FROM HERE, YOU CAN'T SEE PARIS

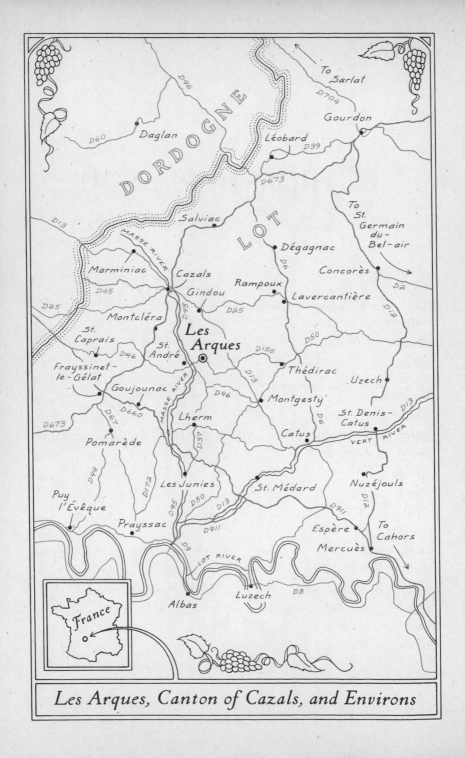

Les Arques, Canton of Cazals, and Environs

INTRODUCTION

On the Michelin map #79, the village of Les Arques, lost in a forgotten corner of southwest France, is the smallest of dots, a speck connected to other, slightly larger specks by the thinnest of lines, denoting the one-lane roads in and out of it. As you approach Les Arques from the south, the passing countryside tempts the eye—herds of goats and sheep, odd bits of field hacked out of the difficult terrain and filled with corn, serried ranks of vines, and drooping sunflowers. Generous farmhouses and barns peer out from far off the road, all built of *la pierre blanche du Lot,* the local limestone, their hand-hewn blocks weathered ocher and bisque under tiled roofs splotched with lichen.

Through the trees on the left, you catch a tantalizing glimpse of the slate-roofed twin turrets of the diminutive ruin of Domaine de Ladoux, then cross a stone bridge through a pasture filled with grazing cows. At last, you begin to climb the hill on which the village is perched like a crown. From below, you see the tower of the twelfth-century Romanesque church at its center, the burnt sienna of the tile roofs, the encircling walls rising up, as if to keep the village from plunging down the steep slope on all sides.

The road skirts the village so that, as you top the rise, the old church and its surrounding houses beckon from the left, while on the right a single imposing structure stands out, an old schoolhouse, and behind it rolling pasture, cornfields, and forest until the eye meets the next ridge at the near horizon.

On an evening in August, the stones still throwing heat from the day, the gravel square in front of the modest town hall opposite the church is filled with people. At first glance, one sees the sunburned faces and work-hardened hands, the suspenders and flower-print dresses, the kids underfoot and young parents pushing baby carriages. Monsieur le Maire shakes hands and gives bisous—busses cheeks—as is the mayor's job on such occasions, while octo- and nonagenarians hold seats of honor on the few benches set around the perimeter. A temporary stage is set up at one end, and recorded old-timey accordion music fills the air.

It is La Fête des Arques, and more than a hundred and fifty people are gathered for the yearly shared meal that marks the start of three days of a festival traditionally celebrated around the saint's day of St. Laurent, patron of village and church. Under a makeshift tent at one end of the square, wooden tables have been set end to end in three long rows, each sprouting bottles of local red and rosé wine at regular intervals and most chalked with one of the local family names. Laval and Bousquet; Tanis, Cantagrel, and Pierasco; Maury, Rajaud, and Marti; Vielcazal and Lafargue—these are the same names carved on the stone sarcophagi in the cemetery on the far edge of town, the same names etched into the stone of the war memorial at its entrance.

To the unknowing outsider, the scene very much fits a mental image built up from television travelogues and the colorful photo spreads of glossy magazines and tour books: three and four generations of family still living in close proximity, farmers and people who work outdoors, under the sky, gathering in the deepening dusk of a medieval village whose meanest pigeon coop appears older than any building in the New World. And what awaits them is a fine communal meal of local food, the aroma of which has begun to waft lightly over the crowd, shifting them in small groups to their places at table.

In times past, the meal would have run to many hours and

many courses, the women of the village competing to outdo one
another with simple local dishes such as goat cheese or onion tarts,
cabbage stuffed with pork and duck liver, duck pie, lamb stew,
rabbit with walnuts, and jam or walnut tarts to finish. All the people participated, contributing their own labor, but also the produce
from their fields, the meat from their animals, and the wine from
their cellars to make the meal.

As my neighbor, Alain Laval, one of the mayor's two sons,
would point out later while we consumed cold salads and couscous
with pot-au-feu, the only things remotely traditional about this
meal were the rough local wines, the bread, and the sugar-covered
beignets served with the coffee. "Papa Couscous" was catering this
meal at eighty francs a head, and the villagers contributed, their
francs and appetites and enthusiasm aside, only in bringing along
their own cutlery, glasses, and china.

The turnout seems quite large for such a small village, one that
has, after all, fewer than fifty houses, and whose entire listing in
the regional phone book takes up less than a single full column on
a five-column page. Start to talk to the festival-goers, and you
soon learn that the kids and their young parents, for the most part,
do not live in Les Arques, but migrate in July and August from
Bordeaux, Toulouse, and the larger northern cities to visit their
middle-aged parents and even older grandparents. The year-round
population of the village is in fact about thirty people, almost
entirely elderly and retired; of the commune as a whole—Les
Arques and its surrounding hectares—169 souls all told.

Thus the fête is catered for the simple reason that the village
women are few in number and getting on in years, and why go to
the trouble for so many birds of passage? Instead they have chosen
to keep what parts of the tradition they can and, for the rest, to
adapt to the different conditions of modern village life. This is but
one of the many changes that have overtaken the village, and only
the most obvious, at that.

In aspect Les Arques seems the postcard-perfect embodiment
of everything our imagination tells us a small village in the south

of France should be. In short, it immediately and hopelessly seduces the eye. On an early morning stroll in summer, the impression is of order and quiet and old, old stones: the cooing of pigeons, a single barking dog, the murmur of old ladies in house-dresses having coffee behind open windows of neatly kept houses that face the street, their shutters spread wide to let in the cool air before the heat of the day arrives.

Flowers sprout from every nook and crevice, shoot up from the base of every wall, burst in profusion from flowerpots and window boxes and hanging planters. Around every corner, down every lane, the eye and nose are drawn by fecund flora of every hue and shape.

And yet this abundance, like the bustling fête, is deceptive. Visitors on a weekend in August might spend an afternoon walking the quaint streets and visiting the imposing but plain church and the small museum that faces it, perhaps have dinner in the restaurant that is the town's sole commercial establishment and then make their way up the hill for the dancing and fireworks later that night in the town square. They would likely come away with an enthusi-astic—and wholly false—impression that Les Arques and by exten-sion the Lot, the *département* (a political entity like a state in the United States, only generally much smaller in size) in which it is located, is a thriving region, one that has managed to keep its medieval character and country rhythm while somehow meeting the challenges of the modern era. The truth is, as always, a lot more complicated and interesting than surface appearances, and Les Arques the exception in a landscape strewn with depopulated vil-lages hanging on by their fingernails to any semblance of vibrancy.

As we finished our salads and began to dig into the generous plates of couscous and pot-au-feu, I caught the eye of Madame Claudine Delon at the next table over. She was one of the first acquaintances I had made in the area when I had first visited a year earlier. We exchanged the ritual *Bon soir!* and *Comment allez-vous?*, commented on the heat and how welcome a bit of rain would be for the gardens and trees.

Madame Delon is a tireless volunteer (and former president) at the Office de Tourisme in Cazals, the largest of the nine area villages, which together make up the canton. She is a key organizer of numerous events from the Week of the Book to the Mushroom Fair, the Dog Show to the Vide Grenier Yard Sale, and also wife of the local plumber. A short, stout woman of cheery disposition and high energy, with a round face and dark eyes framed by short brown hair that never seems to be wholly in place, she knows—and is known to—everybody in the whole canton. If there are a handful who might call her a *commère*, a gossip, many refer to her only half jokingly as "the encyclopedia of the Bourianc," especially the newcomers who have availed themselves of her expertise in calling out an electrician on a Sunday night or finding a room for last-minute guests in August's high season.

When I had asked her so many months earlier about the recent history of the village, Madame Delon had sighed deeply. She talked generally about how the entire region had suffered from war and unfavorable agricultural and economic conditions over the last century, then drew a book from her bag, a well-worn copy of the 1900 census of the area. Perching a pair of somewhat incongruous high-fashion metallic-rimmed bifocals on her generous nose, she began to quote figures.

"Les Arques in 1900," she told me, "had nearly eight hundred people, today less than two hundred. Then, 120 were of school age, today not even twenty. Then, the village had four working mills, two cabarets [small bars], and a brickworks. Cazals had four hotels, eight cabarets, and six cafés! And have you seen St. Caprais?" This is the smallest, most isolated of the commune's villages. "The village is in ruins!" she exclaimed. *"On dirait qu'il y a eu une guerre!"* You would think there had been a war there.

I had been to St. Caprais, another hilltop village with a church at its center, but this one a place whose outer streets tended toward crumbling walls, overgrown bits of field, and fallen roofs, rusting farm equipment spilling out of open-fronted barns and sheds. In the heat of an eerily silent August afternoon, every window shut tered, the narrow, dusty, potholed main street populated only by chickens and scruffy farm dogs, one felt transported back to an earlier but not necessarily gentler era, one of hardscrabble rural poverty, backyard privies, and social isolation. If, perhaps, Madame Delon had exaggerated the degree of St. Caprais' dilapidation, her words struck home all the same. Caught in a downward spiral of eroding population and declining agriculture, it showed what time and the elements will do, in the space of just two generations, to stone and tile without the attentions of the hand of man. Without trade. Without, in short, life.

Only ten kilometers and a single limestone ridge separate Les Arques and St. Caprais, yet they are a world apart: one looking back, a last fond, if somewhat worn, glimpse of the nineteenth century; the other stepping gingerly into the twenty-first century with all its attendant compromise, contradiction, and potentially dislocating change. Even a dozen years ago, it wasn't at all certain that Les Arques would pull itself out of what seemed an irreversible decline.

The village owes its present renaissance to an unlikely duo of phenomena that the French clasp equally tightly to their collective bosom: *le patrimoine* and *la bonne chère*—cultural heritage and good eating. Simply put, Les Arques has a museum and a restaurant. Ossip Zadkine was an émigré Russian artist, contemporary of Picasso and Braque and Malraux, who in the thirties bought a house and barn on the southern edge of the village, where he lived and worked the six months of the year he was not in Paris, until his death in 1967. Perhaps best known for the war memorial sculpture gracing Rotterdam's central square, the *Monument to a Destroyed City*, he is the village's single claim to national fame in the famous sons and daughters category. His legacy, shorn of its less pleasant aspects (he was after all Russian, an urbanite, a modern artist, and born a Jew in a land of rural, Catholic farmers) became a natural focal point when a small group of young people first got together in 1988 to form Les Ateliers des Arques.

Why not invite a handful of artists to come live and work in the village for the summer months? they asked. Why not a small museum? The Regional Cultural Affairs Ministry agreed to fund the projects, and the spirit of Zadkine was so thoroughly invoked that the first simple, one-room museum to his memory opened in his former studio a year later. The visiting artists of Les Ateliers became, over the years, an institution, while the museum grew to three rooms, its collection, borrowed wholly from museums in Paris, overflowing into the courtyard (and even the crypt) of the church across the street.

Artists tend to be thirsty individuals, and they and their entourages need to eat, of course, which is how the schoolhouse, which had closed its doors on its last students for good in 1986, became a restaurant. Originally it was leased to a succession of chefs who failed, one after the other, to make a go of it. Finally — and disastrously, according to Mayor Laval — it was run by the villagers themselves. Mostly, it lay empty for months every year from 1988 to 1994. After six years of losing money and reaping nothing but frustration for their efforts, the villagers were about to give up.

Enter chef Jacques Ratier and his wife, Noëlle, who fall in love with the stone building and its high-ceilinged airy rooms at first sight, who have a vision of an intimate, firelit dining room in the fall and winter, of a sun-soaked terrace of tables in spring and summer in the graveled playground out front. Today, the restaurant, La Récréation — "recess," is booked the entire summer and fall, brings a steady stream of visitors to the village and its museum, and has become a must-see destination in its own right. More unusually, at the same time that La Récréation attracts the summer crowd of vacationing foreigners and French people returning home for August, it has also been adopted by the villagers, hosting the Hunters' Supper and the Golden Agers' Banquet in the fall, the birthday parties, weddings, and anniversaries of the denizens of Les Arques and the surrounding towns large and small.

It was the restaurant that had, in fact, first brought me to Les Arques. My idea was a simple one: to spend a year in and around a small chef-owned restaurant in a rural setting somewhere in southwest France. I wanted to capture the life of a restaurant through the wax and wane of the seasons, taking divergent paths to explore in some depth what made the region unusual along the way. What I found was a far broader, more nuanced picture of a restaurant that had not only played a large part in saving a village but in so doing had come to be in many ways its focal point, the restaurant at the center of this very small world.

The more I began to explore the story of La Récréation, the more I realized how intertwined it was with that of the village and its people, and how both in turn reflected far deeper and quite powerful changes going on throughout rural France and Europe as a whole. In Les Arques and the *département* of the Lot surrounding it, I found a lonely piece of France caught squarely in the headlights of the global economy, the real effects of forces not abstract at all, but writ large on the faces of its inhabitants.

There is Bernard Bousquet, for example, a farmer of about fifty years who lives right next to the restaurant, but who is most often glimpsed in passing, simply because he never stops working. The family farm, La Brugue, where his mother still lives, is three kilometers down the hill. He grows some corn, barley, and oats, sunflowers, and sometimes a little tobacco, but also cuts and sells firewood, tends a plantation of poplar trees, fattens three thousand ducks for a foie gras producer, and rents out an empty village house in the summer. He works an acre of vegetables and herbs for Jacques' kitchen, grows white asparagus in the spring, and also puts in time at the largest local château, feeding the herd of fallow deer that the Parisians who own it have imported.

It has been some time since Bernard could hope to earn a decent living from his 125 acres alone, much less support his wife, both their widowed mothers, and, until they grew up, his own children. And yet they refuse to give up, the Bousquets, improvising new ways to exist out of old, waiting for the tide to turn once more in their favor—which it very well may. At the very least, rumors of the death of this area and of the country traditions that have sustained it are very much exaggerated.

As a stranger here, what had first caught my eye were the lovely contradictions so thick on the ground. Bereft of any major industry and with only marginally profitable agricultural lands, the Lot is France's Mississippi, one of the poorest *départements* in the country. Yet it is known far and wide for foie gras and duck breasts, farm-raised lamb, and saffron, and particularly for its fresh *cabécou* goat cheese, black truffles, and *vin de Cahors*—long-lived very dark wines. Most significantly for the average Frenchman, the Lot, in the year 2000, held the signal distinction of having the most starred Michelin restaurants per capita of any *département* in France, seven in all.

Though this is the land of the Three Musketeers, of the prehistoric cave paintings at Lascaux, of Simon de Montfort, Richard the Lion-Hearted (who never did learn English) and the Hundred Years' War, when you tell a Frenchman you've been here, all he wants to know is which restaurants you chose, what you ate, and which wines accompanied the meal. "They eat well in the Lot," I have heard people from other parts of the country say more than once. It is true, for the Lot is still part of that France in which it is hard to find a bad meal and even harder to find a bad meal at an exorbitant price.

While there is no denying that the old, purely agricultural way of life is disappearing in places and fast-changing in others, more interesting to me was to discover how much of the traditional still exists here and in what guises, how the resourceful younger generation has found ways to adapt and bring into the present what they judge worth saving while the older generation soldiers on, seemingly immortal, inspiring and helping the young by doing what they always have.

Walk by the Bousquet farmhouse on almost any Sunday afternoon out of harvest season, and Bernard will appear on the steps, inviting you in *pour boire un coup.* Sitting in the kitchen with your hand around a glass of muscat or perhaps his own plum eau-de-vie, you will find people of all ages. They talk of the weather, the hunting, how bad the corn was, so-and-so's tractor accident, who is killing a pig next week. This is where the knowledge is passed, where the young absorb the wisdom of the elders, where the local history and mythology are transmitted from generation to generation. It is a very hopeful place, a place where past and future sit side by side, looking, for all we've been told to the contrary, more similar than different. And, happily, Bernard shows little inclination to retire just yet.

ONE MORNING IN AUGUST

By eight o'clock on a late summer day what passes for the morning bustle in Les Arques is well under way. Elise Ségol, the widow who owns the little house across the street from the restaurant, has thrown open her windows and doors for the fresh air. Wearing a voluminous flowered apron over her dress, she leans on her broom, chatting with Max, the gruff village gardener and jack-of-all-trades, who is watering the flower beds around the statue of the Virgin and the war memorial that mark the entrance to the village.

The door of the restaurant opens, and Nougat, Jacques and Noëlle Ratier's rather dim yellow Lab, bounds out and through the gate for his morning rounds. A moment later, an upstairs window opens and Noëlle drapes a quilt over the sill for airing. Inside, the espresso machine whirs and wheezes, and Jacques appears in the doorway, still clad in a tattered blue bathrobe and with plaid slippers on his feet. He has a cup of espresso in one hand, and the first cigarette of the day in the other. He strolls out into the walled courtyard and looks at the clear blue sky, at the white plastic tables on the gravel, then into the restaurant kitchen through the tall windows. His face is contemplative, as if he's measuring the contents of the restaurant's walk-in refrigerator against the hungry crowds who will soon descend for lunch and again at dinner in such lovely weather.

Next door, Bernard Bousquet and his wife, Jeannine, have been up for hours. Their day started at six, when they went down to the farmstead, La Brugue, to feed and water the ducks and pigs and their daughter's horses. The ducks are not a hobby: there are a thousand of them, delivered young to his farm three times a year for fattening before they go to a *gaveur* for force-feeding on their way to becoming foie gras. An hour later, he was home again, out in the back garden. As he does every morning from June through October, he was cutting the brilliant orange squash blossoms that appear, stuffed with scallop or chanterelle mousse and then steamed, as a side dish accompanying every main course at Jacques' restaurant.

Bernard and his wife are sitting around the kitchen table, having a *casse-croûte*, a crust of bread and cheese and dry sausage washed down with a glass of red wine out of a barrel in his cellar. Soon he'll be off on his tractor, back to the fields. At this time of year he has a fortnight's lull; the hay he cut and rolled in July, and the barley and wheat ripen in September. Today he's off to cut the cordwood, mostly scrub oak and second-growth chestnut, which he sells to his neighbors, many of whose houses are still heated by stoves or fireplaces.

Climbing the main street toward the church there are a few signs of life, a cat sunning itself on a balcony, the smell of someone's lunchtime soup simmering on the back of the stove, curtains fluttering at an open window. Many of these houses are unoccupied, or occupied only sporadically by renters or grown sons and daughters spread far and wide who return to the homestead for a few weeks in the summer. Monsieur Mémin, a railroad worker who retired to Les Arques a few years ago, is hard at work doing up yet another corner of his yard for a flower bed. He has taken to creating elaborate floral displays, like the huge overturned kettle in one corner (probably once the bottom half of an alembic still used to brew plum eau-de-vie) from whose mouth spills out a wave of bright impatiens across the green of the lawn.

A hundred yards farther along, the narrow main street opens

up into a square, the town hall on the left, the twelfth-century Église St. Laurent on the far side, and the Zadkine Museum on the right. The town hall, a modest, one-story, rectangular stone building, is closed, its shutters pulled tight, the single bit of color a white sign transected by the bold splashes of red and blue of the French national flag. "Honor to our elected officials!" it reads in bold black type, listing underneath the names of the mayor and his ten municipal councilors.

As if mocking the presence of the state, the church rises spectacularly across the square, its belltower soaring above the roofs of the town and its thick walls of dressed stone dwarfing in their massive solidity the more modest houses around it. Its form is unusually coarse even for a Romanesque church (it was begun at the end of the eleventh century), roughly square with twenty-foot-tall carved wooden doors giving on to the nave at one end and the three semicircular protuberances of the chevet at the other.

In four hours, Madame Jeannine Pierasco will take a five-minute break from her job as a docent at the museum, cross the gravel court to slip inside the church, and prepare to take up her ancestral duty. Like her father and mother before her, she is the unofficial bellringer, using every pound of her frame, hauling down on the rope and then releasing it, to set the bell to rocking and, finally, to ringing out the noon hour. It is a comforting, if somewhat forlorn reminder of a time just a generation or two ago, when village churches like this one had an important place in the daily lives of the inhabitants, when the bells rang out morning, noon, and night to send the men off into the fields, signal their supper, and then ring them home at the end of the day. One perhaps apocryphal bit of folklore has it that villages close to one another would stagger their ringing so as not to confuse the schedules of the locals.

Here in Les Arques, the Église St. Laurent is in no danger of complete abandonment. A traveling priest still says Mass and gives Communion one Sunday a month to the score of older pious who occupy the pews. The interior of the church has a banner here, a carpet there, brochures and a poor box. But one can't help feeling that the major attractions are otherwise. Standing in the doorway, the visitor sees, beyond the pews and directly in front of the altar, a staircase descending to the crypt. There, lit by yellow light reflecting harshly from the pristine white stone, a startlingly Cubist Pièta almost fills the space. Upon leaving, the eyes are equally arrested by another rawly modern work, this one a brutal crucifixion, sprawling in space over the doors.

There was some controversy upon the installation of these very contemporary sculptures, some heated objections from the older villagers especially, even a petition. "A church is a church, not an art museum," an octogenarian who lives just outside Les Arques told me. "Zadkine is all fine and good in the museum, if you like that kind of thing," she finished, leaving no doubt about her own opinion of his work.

Other, more practical, minds, however, see it as no more than the church rendering unto Caesar. The church owes its pristine restoration partly to the intervention of the creator of both of these works, this Zadkine, with the then Minister of Culture André Malraux. Zadkine was a Russian emigré sculptor who settled in Paris as a young man before World War I. While part of the constellation of between-the-wars artists that included Modigliani, Soutine, Picasso, Léger, Foujita, and others of what was then called *l'école moderne*, it appears that his star shone a little more dimly than others, perhaps due to his own modest and retiring ways. Certainly, deciding to spend six months a year in what Parisians still consider the provincial backwater of Les Arques, to which Zadkine retreated regularly beginning in 1935, was not what would be considered today a wise maneuver for an artist whose star was hardly ascendant.

What is clear to all by now is that Zadkine has become an institution in his own right, incarnate in the twelve-year-old museum that bears his name, and that beckons from a dozen feet across the gravel from the church. Zadkine and the church are two cultural icons whose fates have become inextricably linked, for better or worse, with the death of the former, it seems, having bound them together even more tightly.

At the museum, Lionel Gramon shows up early for work. Although it is at least an hour before opening, he will prepare the day's tickets, check the security cameras and environmental control equipment, and make sure the video player endlessly evoking the Life of Zadkine is not having another episode of crankiness. It is a good job, a job he loves, a place that feeds his own, more elusive dream of opening another small museum here, this one dedicated to the everyday life of Les Arques in the nineteenth and twentieth centuries. If only there were a single house for sale he could afford!

Around the corner, eighty-seven-year-old Blanche Delfort has taken up her perch on a chair in the minuscule courtyard of her house. She has been up early, too, picking lettuce, fresh favas, and green beans thin as shoestrings in the garden out back. Flowering plants crowd every flat surface, and the ground is wet from her morning watering. She is waiting, a little anxiously, for her daughter, Simone, who drives in from Agen to spend the weekends with her mother when she can.

Even an hour later, with the sun beating down in a cloudless sky, the houses will be shuttered and their inhabitants invisible, the streets empty but for the passing of the tractor and the mailman in his bile-yellow and blue mailcar. The previous mailman, I would learn, was rather famous locally for the toll his driving had taken on the mail fleet. At one point a few years ago, after he had crashed three cars in as many months, he was apparently restricted to deliveries from a bicycle in nearby Cazals.

Closer to noon, the municipal parking lot behind Elise's house will begin to fill with campers and minivans and the sedans of tourists come to visit the church and museum. The lot beside the restaurant will begin to fill, at first with a trickle and then a rush of Mercedes and BMWs and Volvos with German, Dutch, British, and Parisian number plates. The streets, while not exactly teeming, will see the passage of young families with small children, the middle-aged retired with vacation homes in the area come to stroll and eat, small groups of three generations holidaying together, everyone moving at the pace of Grandma. Most will find their way eventually to La Récréation, Jacques and Noëlle's restaurant, filling its courtyard with the happy noise of appetites being appeased as the smell of lunch drifts up into the rest of the village.

When I moved to Les Arques with my wife, Amy, our six-year-old daughter, Lily, and our black Lab, Isabelle, we did not know a soul. I had spent perhaps two and a half days with Jacques and Noëlle almost a year earlier when scouting chefs and restaurants for this book, had done what background reading I could. We chose our house because, about a kilometer down the hill at the end of the Auricoste road, it was the only one available near the village. Jacques and Noëlle had sent us a video of it, and it took but a few conversations with Gérard Laval, its owner (and the oldest of the mayor's two sons) to conclude the deal. These very tenuous contacts aside, we were entering foreign territory. Elise, Max, Madame Delfort, the Bousquets, the mayor and his family, the restaurant, church, and museum—these were people and places glimpsed once or twice from afar.

We arrived in Les Arques at the start of July 2000. It was a Saturday night, and Jacques and Noëlle had saved us a table at the restaurant. They made a big—and welcome—fuss over us, as much as they could given the fact that they were in the middle of feeding eighty people a five-course meal. Our landlord was out of town, and they had arranged for his father, Raymond Laval, to come get us and take us down to La Forge, which is the name of the house at the foot of the hill.

Monsieur le Maire, short and slight, with work-roughened hands and an accent to match, showed us where the lights were and how to work the stove, delivering the information with the shy smiles and awkward pauses of someone who just does not talk that much. As he rattled off in a creaky white Renault LeCar with much-patched bodywork, I had no inkling of the large, wholly positive role he and his wife and their extended family would come to play in our lives.

The next day, our first in France not spent in a daze trying to figure out traffic signs or flailing about as we searched for one word or another in our sadly dilapidated French, was Sunday, market day in the neighboring village of Cazals. Les Arques, our village, lovely medieval church and timeless stone houses aside, has nothing in it except Jacques' restaurant and a very small museum. No post office, no bakery, no functioning commerce at all. When we first asked if there was a store, most of the villagers would say yes, the small grocery Bernard Bousquet's mother ran right next to the schoolhouse. And then they would add, almost as an afterthought, that it had closed in the eighties. Change takes a long time to sink in here, which is also why, whenever I identified myself as the writer come to write about the village and La Récréation, they looked momentarily puzzled. Oh, the schoolhouse, they would exclaim, yes, it's a very good restaurant!

Those six kilometers from Les Arques to Cazals, along the side of the trickle known as the River Masse, were at first so strange and exotic, the pencil-straight stands of poplar trees in the river bottom on one side of the road alternating with poor farm-houses, flocks of sheep and herds of milk cows grazing green pastures, and spectacular panoramas of the rising hills thick with scrub oak and chestnut and pine. Small tracks sidling off into nowhere intersect the main road, marked by white posts with a single red ring around them, and we caught glimpses of the few grander *maisons de famille* and farmsteads tucked back among the fields, many with the distinctive two-story, square stone *pigeon-niers*, or dovecotes, which feature so prominently in this region.

These large family homes are no longer inhabited year-round, a phenomenon of recent generations as the young have left to find work elsewhere, but have become a holiday gathering place for relatives scattered far and wide, in the summer months especially. The fact that there were cars on the roads and in front of every house, that people were walking the less-traveled lanes, led us to believe, at first, that we had been greatly deceived about the isolation of this part of France.

After five kilometers of tight turns and switchbacks, the landscape opens out into flatter country, and the road becomes an allée planted on both sides with forty-foot-tall, thick-trunked, immaculately pruned plane trees. The French can't seem to resist taking the shears to anything that grows, which sometimes has the effect, in the more inhabited areas, of making all the roadside trees appear to have the same, quite elegant haircut. The houses, of stone or clay block stuccoed in earth tones, come closer together, and you begin to notice the exquisite care people take with their flowers and shrubs here, the bright blooms particularly set off by the softer tones of gray, brown, and white of the ubiquitous stone. Finally, the last hundred meters take you past stubby pollarded locusts and walled gardens from which sprays of delicate roses and tendrils of wisteria and jasmine escape until the road intersects the main street at the Cazals town hall.

If you stop right here, take ten giant steps to the right, and look skyward, you can see the château crowning the hill above the older section of town. Through half-closed eyes blocking out the newer parts down below and to the left, you can just make out (especially if you've got the brochure from the Office de Tourisme in front of you) that the houses above were built in tight rings around the base of the château, the whole once enclosed by a massive encircling wall whose stones have been long since pillaged to build the rest of the town.

Cazals, at eight hundred residents, is the largest of the group of nine villages (Marminiac, Cazals, Gindou, St. Caprais, Frayssinet-le-Gélat, Montcléra, Goujounac, Pomarède, and Les Arques), which together make up a governmental unit known as a *canton*, roughly equivalent in size to a (modest) American county, with Cazals as its seat. This canton is called the canton of Cazals, and is part a larger governmental district called Sud Bouriane, or Bouriane South. Bouriane is an old word in occitan, the patois they have spoken here forever. It means land of the mule, an apt name for a landscape of steep hills and rocky paths in which most transport was by mule.

Cazals was originally a *bastide,* or fortified "new" town built by the English in the twelfth and thirteenth centuries during the Hundred Years' War and is not a place distinguished by its château, which lurks behind a high fence and is privately owned, nor its church, your basic eighteenth-century hilltop pile. The church, which replaced a far older structure that burned, is new as churches go around here, its architecture blocky and graceless. Cazals' true identity is as a lively commercial center that, even as recently as a few short decades ago, when the area was much more heavily populated and agriculture was king, was known for its monthly livestock sales. Today the town flourishes in a quiet sort of way because of its *petits commerces,* its small businesses. There are two bakers, two butchers, two small groceries, two cafés, two restaurants, one hotel, a newsstand/tobacco shop, a hairdresser, a dentist, a doctor, a taxi service, an appliance store, a hardware, a TV repair, a plumber, an electrician, a gas station, a garden shop, and a clothing store. There is also a post office, the Trésorie Municipale (to pay taxes and your child's school lunch bill), the municipal day care, and an elementary school.

And, of course, on Sunday mornings from seven to noon, there is the market. In the beginning it was the market that began to introduce us to how the French live, eat, and socialize. In Cazals this institution occupies the town's center, Place Hugues Salel, in summer swelling to occupy the whole square with three rows of stalls large and small set up early in the morning on what is otherwise an asphalt parking lot shaded by young plane trees and surrounded by two- and three-story seventeenth- and eighteenth-century buildings whose ground floors are given over to those other small commerces.

Certainly, the market is a feast for the senses, especially in high summer, when so many more farmers come with their wares. My notebooks from this period are filled with embarrassingly complex and intricate descriptions of fragrant fist-sized Quercy melons and the tang of fresh *cabécou* goat cheeses; of gamy duck and boar sausages and freshly butchered yellow-skinned farm chickens; of apricots, strawberries, plums, grapes, and other produce. The sad fact, and it does bear repeating, is that the lowliest carrot or head of lettuce here is of a freshness and flavor that makes even the best American supermarket products seem pale imitations.

Madame Stuker sells more than a hundred kinds of fruits, vegetables, and fresh herbs at her long, shaded stand, which has a prime position in the middle of the first row. Her offerings change with the season, a riot of fruits in the summer seen at no other time of year, for example, which gives way as the leaves fall to a rich selection of root vegetables and a dozen varieties each of potatoes and cabbage, squashes and hearty lettuces in winter. This accurately reflects what is available very locally, within perhaps a hundred miles, which in turn reflects what people here are used to eating. Morrocan clementines and bananas from the Carribbean in the winter are the exception rather than the rule, with most people preferring to choose from among a wide offering of pears and apples.

Each wooden crate bears a chalked sign marking not only the price, but the provenance of the item. This is not Madame Stuker's conceit but what her customers demand. No one would imagine making a winter cassoulet with any but the white beans of Tarbes that appear in late fall. And who would garnish their summer salad with Dutch hydroponic vine tomatoes when the beauties from Lot-et-Garonne were harvested yesterday afternoon fifty miles away?

While her husband makes the rounds of the farmers and orchardmen and wholesalers who supply them, Madame Stuker travels five days a week to different markets in the region in her refrigerated truck. About forty years old, thin and fast-moving, she is what the French call *une jolie laide*, literally, a comely, ugly woman—the kind of woman whose features, though not traditionally beautiful, add up, with the spark of her personality and easy smile, to an attractive whole.

She took it upon herself to correct my mistakes—not only in vegetables but in French as well—gently but insistently. Soup carrots, the big ones, are called *les grosses* and not *les grandes*; one can ask for *une dizaine* or *une douzaine*, about ten or about twelve, of something, but not for about five or six. Ask for *celeri* and you'll be handed a round, yellowish-white root the size of a grapefruit: celeriac. (What you want is *une branche de céleri*.) And she was ever ready with rapid-fire instructions on how to care for what she sold and when to eat it. "The pears won't wait, and those fingerling potatoes are not for the oven! And the avocado is for tomorrow, *hein*?"

The Cazals market also has, more typically, at least a dozen creaky old farmers, men and women, who sit in lawn chairs under umbrellas, a couple hundred fresh eggs or a few dozen tiny goat cheeses or twenty lettuces and six bunches of carrots on crates beside them. It has three local duck farmers who sell their own foie gras and preserved duck pieces (confit) and duck pâté and other delicacies. It has a tin man, selling mostly junky kitchen utensils and plastic containers and cheap pots and pans. There is the wineseller, pushing second-quality red *vin de Cahors*, rosé, and Rivesaltes sweet wine in bottles and in *vrac*, bulk, where you bring your own five-liter plastic jug to be filled with a hose from the tanks in the back of his truck. The traveling shoe store comes and goes depending on the season, as do the oyster vendor, the flower sellers, the olive cart, the jam lady, a trio of organic farmers, and one woman who sells the deadly local homebrews, *vieille prune* and *vieille poire*, aged plum and pear eau-de-vie that can take the paint off the side of a house.

One of the most likable, and certainly the most theatrical, of the vendors is the cheesemonger, who, in the summer, dons a wide straw hat and puffy-sleeved black peasant shirt and flirts with all the ladies from behind his piles of mountain cheeses, hard Cantal and ewe's milk and blue cheeses, with an occasional Brie or Camembert thrown in for good measure. The market also has a Silent Sam who hovers, gaunt and austere, behind his meticulously stacked and neatly labeled bags of the three local varieties of dried mushroom (porcini, morels, and girolles), and a charcutier from the neighboring *département* of the Avéyron who extolls to passersby the excellence of his dried smoked sausages and hams and is always handing out samples.

In the summer, you have to get up early if you want to see the people who live here. By eight A.M., small clusters of the middle-aged and the old gather between the stalls, or work their way from one end to the other, the men in best cloth caps and clean coveralls or Sunday suits, the women in voluminous flower-print dresses and straw hats. *Bisous*, the traditional kisses of greeting on both cheeks, are exchanged and hands shaken. Rolling papers and a tin come out, and they settle back, cigarette bobbing in the corner of the mouth, for a palaver. They chat up the sellers, often their friends and neighbors, usually in patois, genially deriding what is on offer before making a few carefully considered purchases, which disappear into enormous cloth bags or straw market baskets, then meander down to the café for a glass of pastis or sweet muscat or beer or an espresso.

An hour later, they have vanished, puttering off in their diminutive and much-abused Citroen deux-chevaux and Dianes and LeCars, all belonging to that subspecies of French auto cheap to buy and cheaply made, seemingly designed for dwarfs, and powered by a tiny, noisy engine unencumbered by antipollution devices consuming minute amounts of leaded gasoline or diesel and spewing particularly foul fumes.

By ten the lanes between the stalls are packed with a different crowd, the summer residents, who have stirred only recently from their beds and who have begun to consider what they will eat for dinner. They arrive in minivans (which really are quite mini here) and station wagons, kids in tow, Grandma and Grandpa sometimes, too. They are immediately recognizable if for no other reason than that some wear shorts and most, especially the Northern Europeans, are sunburned. The Lotois, those of all but the very youngest generations, subscribe very definitely to the kind of country modesty that precludes the wearing of shorts but not apparently relieving oneself two feet off the main road.

There are significant numbers of Dutch and Germans and Belgians, but the vast majority seem to be sun-seeking Brits, who most recently discovered the Lot a decade ago and have been buying up summer homes ever since. They have a prickly but workable relationship with the Lotois based mostly on the fact that the English spend wads of money, rent surplus houses empty the rest of the year for inflated prices, and employ local masons, electricians, and roofers to fix up their vacation homes.

The other pleasant discovery we made those first weeks was the too-grandly named Café de Paris, on one corner of the market square. Outside, there are a dozen or so white plastic tables, on market day packed with people. Isabelle and Sandrine, the two young waitresses, scurry about all morning with trays of espressos and café crèmes and an equal number of glasses of beer and Ricard, while the octogenarian owners, the usually dour-faced Monsieur and Madame Bouyssou, look in now and again from their butcher shop next door.

Sitting at a table outside, you can follow the progression of a French family doing their rounds, first through the market rows for produce and flowers, then across the street to the *tabac* for cigarettes and newspapers and Lotto tickets and candy, to the corner butcher for two days' meat, then down the street for some of Bruno Bouygues' excellent bread. They arrive, cloth bags bulging, set down their purchases, distribute the sweets to the kids, and then order minute, deadly strong espressos followed by a glass of something, an *apéro* to get the stomach working before the early afternoon family supper to follow.

The struggles we had our first Sundays were not so much about what food to buy as to how to avoid buying too much, such was the temptation. To take a single example, August is a fruit lover's paradise in the Lot. My daughter has a weakness for strawberries and raspberries, so we always stopped first at the berry lady, asking her to set aside a box of each for us, especially the *maras des bois*, small, cultivated strawberries but with all of the intense flavor of their wild woodland cousins. But then, apricots were also coming into season, as were the Quercy melons and the first figs, Chasselas table grapes, white and yellow peaches and nectarines.

At the charcutier's stall, there are at least forty different home-made offerings from which to choose. There is delicate, almost white boiled ham and dark, salty prosciutto-like *jambon de Bayonne*, a slightly less powerful *jambon de pays*, boar ham, and a score of sausages of all shapes and sizes, fresh and dried and smoked, stuffed with Roquefort or mushrooms, made of beef, pork, duck, boar, cured under ash or in black pepper, perfumed with garlic or other spices. The solution was to try one or two each week, a policy we extended to the cheeseman's wares as well.

Because all the French people we knew before coming had filled our heads with ideas of how empty, remote, and positively backward the Lot is, we were quite surprised to find it packed in the summer with such a cosmopolitan collection of visitors from all around Europe, every restaurant and café full at mealtimes, the roads like as not jammed and the nights noisy with activity. How little we knew!

To the people who live here, the brief swelling of the popula-tion in summer is an aberration, and often a pain in the ass, but a profitable one at that. The campgrounds are full, everybody's empty second house or converted barn is rented for the whole summer, the pool man and satellite television guy work eighteen hours a day, and all the visitors seem to want to do is to buy up everything the locals can grow, make, or import. If you don't actu-ally see much of the people who live here year-round during the day, it's generally because they're too busy working—although they generally keep to themselves anyway.

The Lot in July and August is also quite hot. A dry heat, but one nevertheless that forces you very quickly to become acquainted with some of the ancient verities. In our house, for example, a one-and-a-half story farmhouse with two-foot-thick walls, every window has shutters, in our case painted a lovely and quite common milky shade of blue known as *le bleu gaulois*. They are much more than fashion accessories, however, which we learned from watching our neighbors. In the early morning, when the night air still lingers, people hereabouts throw open windows and shutters to capture its coolness. By about ten o'clock, however, you shut the windows and shutters, effectively turning the house dark inside but creating a welcome haven from the outside, which becomes furnacelike in a matter of hours. (This phenomenon does not, however, explain the fact that most people here keep their shutters closed during the day no matter the season.)

July and August were two very strange and difficult months, long sun-blasted days in an exotic landscape in which everything was new, and our place in this small world uncertain. Our daughter, usually the most accepting and distractible of children, suddenly became afraid of the dark, and of bugs, collapsing into tears and asking quite piteously when we were going back to Maine. I had moments like that, too, especially after a wave of invasions by the natural world, beginning with the ants, then mice, then flies, and then the noisy wasps called *guêpes*, whose sting is very poisonous. (Here, they say the way to chase off a wasp flying around you is to sing softly to it.) They built a nest just above my daughter's bedroom window, and she went to bed each night to the sound of their buzzing.

Having emptied countless cans of bug spray, we finally consulted the mayor. "Why, you must call the *pompiers*, the fire department in Cazals!" he said, as if it were the most obvious thing in the world. One evening at dusk two days after we called, a red fire truck the approximate size of Ford Escort appeared in the drive, ladders on the roof, and two *pompiers* got out. The rotund older guy set up the ladder against the wall at the top of which the wasps had their nest, while the teenager climbed into what looked like a space suit. The feet of the ladder sat precariously on top of another stone wall forming the staircase up to the house. The kid, hardly able to move and now encumbered by a pressure tank and sprayer hose, paused at the bottom and looked a little hesitantly at his colleague.

"Don't worry!" the other man said, putting a hand casually on the ladder rail. "I wouldn't let you fall! Your mother would kill me." That was the end of the wasps. And, for a period of several weeks, my wife's favorite nightly entertainment was watching me hop around, half naked and waving a towel, chasing errant bats out the window. How could a culture that had given us so much, from Molière to Proust, Pasteur to foie gras, have somehow missed the invention of the window screen?

Thrown back on our own resources socially, with no friends, a six-year-old to be entertained and a young black Labrador in need of daily exercise, we found ourselves embarking, in the cool of the morning or late afternoon, on long walks to explore the village and the surrounding countryside. In rural France, we soon discovered, you can do no better than have a bright-eyed child and a sleek, gentle dog as your social ambassadors.

Almost directly across from our house a shaded lane breaks off from the main road, meandering in a lazy S past an acre or two of barley and corn, a walnut grove, over a stream called the Divat, and then past the house and two barns that make up the small farm of Madame Trégoux before joining the larger road up to the village. This lane is marked at either end with the ubiquitous *panneaux*, the small white signs with blue letters giving the *lieux-dits*, or place names, by which every individual locale is identified in this country.

Madame Trégoux's land is called Auricoste, and has been since 1519, when a watermill, the last of five, was built along the Divat by a family of that name. You are never far from history out in this countryside; indeed what Americans consider the past is often here nothing more than the way people live their daily lives. These mills still stand, three as private homes and two abandoned, but all five having continued to turn up well into our time milling flour, grinding animal feed, pressing walnuts for oil, making electricity, and even working the bellows of the iron smelter. This last mill, on the Castelfranc road just below our house, turned to sawing wood when iron became unprofitable around 1875, and its last owner was Madame Trégoux's husband, who kept its wheel turning into the 1950s.

You could never guess at the importance of Auricoste in the past life of Les Arques from looking at it now. The Auricoste road is so little traveled that grass grows down its center, and the stone walls that line it have lost many of their curious, spade-shaped capstones. Auricoste, by which people here mean the whole flat valley on this side of the steep hill on which the village sits, was once in the middle of the richest bottomlands of the parish, and the Divat provided not only irrigation and water for the mills, but also water for drinking and for washing clothes.

This last is the source of communal memories still alive to some of the older residents of Les Arques. Just upstream of the bridge is the site of the village *lavoir*, large flat stones angled toward the stream and lining its banks for one hundred feet on either side. Once a week from time immemorial well into the twentieth century, the women would load up the wagon with dirty clothes and buckets of ash (but never from oak fires, whose ashes stain) and descend in groups to the *lavoir*, there to beat the clothes on the smooth stones—and to exchange gossip and news with others from nearby farms. The narrow path they used still exists, a steeply pitched half-mile slog faced on the uphill side with an almost continuous drystone wall taller than a man. This track shoots from the corner of Madame Trégoux's farm almost straight up, coming out at the top of Madame Delfort's garden behind her house.

Although what drew us to Auricoste first was its serene beauty, we soon began to visit for another reason, to talk with the old woman we had occasionally glimpsed sitting in a chair beside the open door of the farmhouse.

Madame Trégoux, Yvonne, is ninety-three, and she stays in the old house from March until the cold of December, keeping watch over the barns and growing a small garden, but mostly, I suspect, to maintain her independence. Yvonne, at a little more than five feet five inches tall, with few teeth and a face like an apple doll, is at first of fearsome aspect. Her usual expression is a scowl, black eyes sparking fiercely, her usual conversation a catalog of moans, growls, and odd popping and spitting noises emerging from her pursed lips, the whole punctuated by waving of arms, despairing shrugs, and intricate gestures with the hands. These wordless utterances, coming at the beginning, middle, and end of her sentences, express, more often than not, disgust with the lacks of the world and the people around her, her failing body, the cold of night, the rats who are stealing her corn, and the foreigners who make off with her walnuts.

Her house is a traditional eighteenth-century Lotois farmstead, a simple stone rectangle with a clay tile roof perhaps a hundred feet long and thirty feet deep and of two and a half stories. Standing at her front gate and looking at the house, you see the perfect inverted U of a wide Roman arch on the ground floor to the left of a massive stone staircase leading up to the front door, on the first floor. Under the arch are two wooden doors weathered gray, the entry to the ground-floor byre. Oxen and milk cows and pigs were too valuable to leave out in the field at night, and in the winter the warmth they generated filtered up to the family living above.

There is a single, usually shuttered window on either side of the door, and above, just below the roofline, three much smaller openings framed in dressed stone, of an adequate size for small windows, have been let into the wall. Yvonne's windows are diamond-shaped, but they can be oval or almost round, what they call *l'oeil-de-boeuf*, or a bull's-eye. Most every farmhouse here has them in one form or another, for they serve to ventilate the attic, releasing the heat from the house at night but also keeping the wooden beams and lathing on which the heavy clay tiles are set dry and free from rot. The Trégoux family never actually bothered to put glass in them, however, instead boarding them up except for small holes in each corner.

The house, whose walls are wholly of the rough, many-shaped limestone rocks removed from the fields after every plowing, is also a showcase of mason's shortcuts. Auricoste does not sit on a foundation of larger stones set into the ground; instead the walls are thicker at the base, tapering in width as they rise. And the house uses a minimum of the more costly, dressed stone required for the square openings of doors and windows, and yet the effect is a pleasing one. For example, to make those smaller upper window frames self-supporting without the need for a more complicated arch overhead, the stonecutter has taken two rectangular blocks of stone and cut out triangles from one long side of each. Setting the blocks one on top of the other with the edges of the cutouts aligned makes a diamond-shaped window opening of but two blocks of stone. And here, some effort was made to align the windows vertically, such that this small attic window sits directly over the rectangle of the long window below, which is itself centered above the arch of the byre, the variation of shapes pleasing in its simplicity.

Yvonne's house is also typical in its state of selective neglect. The roof is new, as is the zinc guttering at its edges, but the front doorsill has rotted away completely, the window frames and shutters have shed their white paint, and the downspouts end ten feet off the ground, feeding into white plastic irrigation tubing that trails off across the yard. One section of the front garden wall has fallen in, the iron gates ceased long ago to swing from their hinges, and the barn wall facing the house has been repaired with any material to hand—cement block, brick, and even concrete in patches.

I have often stood in front of this house, marveling at the collection of objects strewn about, more or less in neat piles around the house and barns or leaning in orderly rows, as if on display, against their lower walls. There are the old roof tiles quietly decomposing into piles of red dust; seed and fertilizer bags neatly folded against some future need; plastic hose of various lengths; bags of dirt; mounds of rock grown over with lichen; ends of zinc guttering; at least a dozen pails, buckets, and pots of rusting metal, clay, and plastic. Coils of baling twine occupy the empty rabbit hutches to the left of the house, while a twenty-foot-long banister leans against the outside of the house on the far side of the steps behind a long-dead palm tree still shedding its fibrous branches.

The small barn at right angles to the house holds an even more interesting collection, beginning with a Peugeot two-seater so ancient its tires have rotted, a lovely green *deux chevaux*, piled bales of hay, a wooden-wheeled farm cart, trailers of various sizes, three hand-hewn ladders, a water tank on wheels, screens for drying walnuts and plums, plow blades and cutter arms, miles of heavy black plastic, a wheelbarrow entirely of wood, a jumble of pallets, three or four dozen old tires, a harrow, a tiller, an iron plow, and a windrower, as well as two other, unidentifiable pieces of farm equipment all rusted to the identical shade of chocolate, a tractor's engine block, and, bearing pride of place, a massive silver-painted 1955 Rivierre-Casalis hay baler (*70 coups de piston à la minute!*).

Auricoste is one of my favorite places in the whole commune, whether or not anyone is home. When I stopped by on a walk, it was rare that discussion of any one of these items with Yvonne or her daughter, Marie-Josée (still plowing and tilling atop her tractor through well past retirement age) didn't lead to another history lesson on farming or village life here in the past decades—if not centuries.

The first time we met Yvonne, in the early summer, she was puttering around her yard. We stopped to introduce ourselves, and soon discovered that her fierce exterior hid a warm and gentle creature, a person perhaps lonely and out of the habit of talking. The next time we passed, a very hot afternoon, she invited us in for a drink of water. She was fascinated by Lily, our daughter, who was just learning to speak French, and coaxed words out of her with cookies. The first time I saw her smile, really smile, was as she watched Lily play with a biscuit tin, her eyes following the little girl as she explored the kitchen.

"I had to learn French, too," she told Lily. "When I went to school, I spoke only patois, like my parents, what we spoke at home. We didn't know what French was!" she cackled. "And we didn't go so long, either. I left when I was twelve, to go back to work on the farm. That's the way it was." (Patois endures to this day. Picking up my daughter from her school in Marminiac one afternoon months later, I heard a mother complaining that *la maîtresse* had asked both parents to stop speaking patois to their four-year-old, as it wasn't helping the child to learn proper French.)

Yvonne's kitchen, the only room we've been invited to see, could be an exhibit in early nineteenth-century provincial farm life. From the door, you look into a large, square space dominated by a single long table and sturdy caned chairs in its center. Underfoot is rough wooden planking, while the ceilings, smoke-stained plaster between blackened beams, are at least nine feet high. One half of the back wall is taken up by a massive fireplace and chimney, called a *cantou* in this region. The hearth is three feet deep and eight feet long, with a massive beam of chestnut for a lintel at head height and a forged iron swing arm set into the back wall a couple feet off the ground for hanging pots over the coals. The whole is easily tall enough to stand up in once you duck under the lintel, and it had to be, for this is where the family hung the hams and sausages to cure after the fall pig slaughter.

The *cantou* with a chair on either side of the fire was the traditional perch of the elders, and two old folks sitting in the fireplace warming their arthritic bones was what you saw when you walked into any farmhouse in the Lot until the seventies (the 1970s), Suzanne Laval, the mayor's wife, would tell me later.

To the left of the fireplace at hip height is a rough stone sink, the *evier*, a single piece of flat stone six inches deep shaped into a shallow trough, the triangular back end disappearing into the wall of which it is a part. The wastewater runs out the back, draining directly into thin air. These Y-shaped protuberances that jut out of the exterior walls of many old houses at about head level had puzzled me, and now I understood finally what they were. The room was dim, lit meagerly by a single large window above the sink and another thin slit by the door. In one corner, a massive breakfront rose almost to the ceiling. Its two-foot-wide lower doors were made of single planks of chestnut dark with age, the hinges forged iron. Next to it was an equally massive grandfather clock. There

was a vintage gas stove against one wall, and a miniature refrigerator, but these were her only conveniences. Heat was from the fireplace, hot water there was none, and neither did she have a shower,
telephone, television, or even a bathroom, gesturing to the great
outdoors when my daughter was in need.

When we left after that first visit, I rather forwardly took her
hand and kissed her lightly on both cheeks. She cooed with pleasure, and I realized that, like my own grandmother, she had most
likely outlived her contemporaries and had few people in her life.

Yvonne Trégoux—together with her house, her life, and her memories—gave us our first glimpse into a rural lifestyle that most
Americans would find rather rude and certainly very hard. Her
standard of living (she lives on a farmer's pension of about four
hundred dollars a month) represents an extreme even for this little
corner of the southwest, and yet, walking the forest tracks and
tractor paths of Les Arques, one can see that the era from which
she comes is, for better or worse, still very much alive. The majority of houses here do have indoor plumbing and central heating,
but in many things, especially when it comes to doing for oneself,
the country ways persist.

In the village, many people grow kitchen gardens, even on the
tiniest scraps of ground terraced out of the plunging hillside. Two
steps out of the village and you find that, not only do they have a
large garden, but chickens, rabbits, ducks, and perhaps a pig and a
steer or two. They grow a bit of corn and oats for the animals, bring
in the hay in the summer for winter fodder, and, whenever possible,
cultivate a half acre of vines to make their own wine. They have
orchards large and small of walnuts, apples, plums, and pears, and,
although it would be some time before we realized what they were,
even occasional plots of oak and hazelnut in the soil under which
they still look for black truffles in the winter months.

There are 169 year-round inhabitants of Les Arques (fewer
than thirty living in the village proper), only a handful of whom

identify themselves as farmers. Yet, sitting on our terrace in the cool of the evening, so many vintage McCormicks and John Deeres and Massey-Ferguson tractors would come lurching and bumping their way up the steep road past our house that we came to call seven o'clock the hour of the tractor. They were hauling flatbed trailers stacked high with rolls of hay or steep-sided wagons filled with golden oats or barley. Where was it all going? For a place with so few farmers, there certainly seemed to be a lot of cultivated fields, numerous smallholdings, and ample livestock of both two- and four-footed varieties.

Like many summer visitors, we saw only the picturesque in what was cold, hard necessity, only the quaint in traditions that the people here maintained for reasons, very good ones, in fact, that it would take us some time to comprehend. In the beginning, it was all one great puzzle.

TWO

SUMMER IN THE KITCHEN

The pressure to begin my kitchen apprenticeship was something I felt from our first days in Les Arques. The village and its people I reckoned to get to know throughout the course of the year, as our daily encounters rendered me just another harmless neighbor who happened to ask questions all the time. But the life of the restaurant changed with the seasons, and summer was ebbing away. I itched to get into Jacques' kitchen (and his life, but he didn't know that yet), but the daily grind of reality kept intervening. Finally, after weeks spent outfitting our house, buying our first car, opening bank accounts, acquiring French bank cards, and firing off the first shots in what would turn out to be a protracted war to get our residency permits, I could turn my thoughts to what had brought me here originally, Jacques and Noëlle and their restaurant, La Récréation.

La Récréation is not the first restaurant ever to open in Les Arques. Although we can never know for sure, it is likely that there was the medieval equivalent here, a hostelry, attached to the original Benedictine monastery or just outside the village walls, eight hundred or even a thousand years ago.

The hostelry would have served the region's first tourists. These were not landscape gapers or holidaymakers, but religious pilgrims, who, beginning about A.D. 950 began to pass through on the well-trodden route from Vézélay in eastern France and other northern cathedral towns to the southern Pyrenean passes and on to St. Jacques de Compostelle in Spain. The pilgrims traveled in groups for safety, stopping at night in the walled, hilltop villages in which the region abounds. At a time when what little established authority existing outside the major cities was ecclesiastical, it was the monasteries that catered to the needs of such travelers, housing and feeding them and profiting into the bargain with the sale of everything from indulgences to relics.

The scene in the kitchen of La Récréation this August night, great heat, clamor, searing flesh, flaming pans, and demented white-aproned staff careering hither and yon, would more likely move a medieval pilgrim to thoughts of the infernos of hell than of a good dinner to come. To me, from my perch in a corner against the front windows beside the bread and soup station, it was better than the baby Jesus in velvet booties, as the French say, to express sheer joy.

I am not a chef, had not indeed set foot in a restaurant kitchen except by accident since my first employment as a dishwasher in what passed for a fancy restaurant in rural Pennsylvania twenty-five years ago. As someone who finds solace in the kitchen and at the table, I had set off on this pilgrimage with a simple desire to explore the secret life of a restaurant. Combined with a lifelong attachment to France and the French language, fate had led me to Jacques and Noëlle Ratier and to their thriving creation.

In summer at La Récré (lah ray-cray), as it is known for short, people eat outside, either under umbrellas or at tables under the eave of what used to be the long narrow stable that forms one edge of the courtyard. This doubles the seating capacity from 35 to 70, although Jacques in his crazier moments has served up to ninety covers. Which he and his one sous-chef and two *commis* and three waitresses were well on the way to doing before my eyes. (*Commis*, literally assistant, is the lowest rank on the kitchen ladder, and their work runs the gamut from all the basic prep to mopping the floor, to working a station like cold entrée or desserts during service.) Although that number of patrons might seem small to some, serving a five-course meal to ninety people translates into the preparation and delivery of 450 dishes, and that is not including drinks, wine, or coffee. It was going to be a long night.

The afternoon had passed in a blur of the rituals of prep, in which few finished dishes (except desserts) are actually made, but the pieces of their making put into place. Earlier I had watched Jacques reduce a two-foot-long, twenty-pound faux fillet of beef, cut from either side of the back along the backbone, which is less used muscle and therefore more tender, to delicate *pavés*. (Filet mignon comes from just under this long piece.) With deft strokes of a stubby boning knife, he first trimmed away all the fat and fascia from the meat's surface, then stripped out any white, stringy sinews with the tip of his knife. What was left he shaped into one long, roughly square-sided rectangle before slicing into portions. In the process, he discarded about one-fourth of the cut's total weight, but was left with beef so fine that, when cooked rare, you can cut it with a fork.

"I count on about six ounces a serving," he had told me. "But in the summer, I buy five or six faux fillets and throw them in the walk-in and don't give it another thought. They'll keep twenty-one days, but they're gone in a week. In the winter I have to calculate a little more carefully. The pieces like this?" He held up one of the ends of the cut that was a little more irregular, a little more shot through with sinew too deep to slice out without reducing the piece to ribbons. "This I'll save to give to people who order their meat well done."

"The English will eat it." Fred muttered from across the room. Fred, five feet, six inches in his kitchen clogs, with black ponytail and equally black eyes, is Jacques' twenty-three-year-old sous-chef, a diminutive dynamo much given to rude jokes in patois and waitress hassling as he works the stove at Jacques' side during service.

"Well," Jacques added, grinning a little wickedly. "He's right. *Pour les français, ça va pas du tout!* The French order their red meat *bleu* or *saignant*—very rare or bloody—and seeing such a piece on the plate for them would be an outrage."

One by one the hundreds of little tasks that make up dinner prep were accomplished, an hour-and-a-half scurry leading up to the staff meal. At one point Jacques ducked into the walk-in, emerging with a container of chicken livers. A small saucepan had been warming on the stove, and Jacques added butter. As this melted and began to sizzle, he'd already chopped the liver finely, added a dash of salt and spice. As soon as it went into the pan, he began to toss it (*sauter* means to jump), using that particular motion of sliding the pan forward over the flame then giving it a little jerk, which both kept the pieces of liver from sticking and sent each up into the air and down again in a backward arc that keeps them in the pan.

Jacques tasted a piece, nodded, reached for some vegetables—green beans chopped, a little celery? He lowered the flame and grabbed a few slices of stale bread, reducing them to small squares with a few strokes of the knife. Into the pan they went, too, flip, flip, flip.

What was this little something a-creating? There was no liver on the menu that night. Some little chef's secret, perhaps? I began to smile, very happy, feeling that rush of excitement—this is why I had come, to experience things like this, get the kitchen skinny!— my pen skidding over the paper. Jacques, meanwhile, was on to other things.

Finally, he took the liver, vegetables, and bread from the stove, threw it into the top of the industrial Cuisinart chopper, blitzed it once or twice, and threw it into a heavy-based but shallow bowl that looked very familiar, looked in fact like my dog's food dish. Lastly, he thrust a finger into it to test the temperature, nodded, then strode out into the courtyard, whistling for Nougat, his yellow Labrador. Chastened, I put my pen down and took a silent vow never to take anything I saw here for granted. I looked up and Jacques was back, already washing his hands, eyes on the clock, consulting the never-ending list in his head for the next thing that had to be done before prep was up.

A few minutes later, Jacques went to the stove, gave the help's dinner—an enormous skillet of mussels sautéed in butter, cream, white wine, shallots, tarragon, and chervil—a couple of shakes. He took a basket of frites from the fryer, shook the oil from them, and threw them into a bowl along with several lashings of salt. *"On mange!"* he called out, the last word coming out more like mheinzhe than mohnzhe in his deep Toulousian accent. (This is also why he pronounces his name Zha-keh, almost as in the English Jack, instead of the Parisian Zhock.) "We're eating Belgian tonight, kids, *moules-frites*," he said.

Everyone gathered in the dining room, empty of tables, which had all been moved outside under the umbrellas except for a single long one covered by a white cloth. On it was a huge tub of ketchup, carafes of water and red wine, two small Camemberts, a basket of day-old bread, and two six-packs of supermarket crème brûlée and pecan pudding. They stubbed out their cigarettes—except for the English waitress, Jeanette, they all smoke—served themselves quickly, and soon the only sound was that of mussel shells being sucked empty and hitting the plates.

Jacques surveyed his brood from the head of the table, a Mediterranean Paul Newman with cropped dark hair going gray over eyes the color of a clear summer sky, an aquiline nose, and strong, clean-shaven chin. His white short-sleeved chef's tunic is always ironed and immaculate, and, together with his noticeably developed biceps and forearms and shipshape appearance, gives him the look of a haute cuisine master sergeant, an effect only emphasized by his erect posture and commanding voice. "Where's Mireille? MIREILLE! *ON MANGE!*" he called through to the kitchen. She was the summer dishwasher, thorough, but slow. Finally she came in, wiping her hands on her apron.

"So, Mireille," Jacques asked. "You're going to get married in what, two weeks now?"

"Three." She said then, primly: "I still have time to reflect." The table erupted in laughter. Mireille must be seventy years old.

The phone rang. Noëlle picked it up, turned away, said a few words that were obviously not about a reservation, and hung up. She turned to Jacques with a funny look on her face. "That was Anne. She's not coming back. She says she can't stand to work in such a psychologically oppressive atmosphere anymore."

"*Merde!*" said Jacques. "Betrayal!" He looked away, smoking silently, then focused on Noëlle, who had risen from her chair, her face gone completely white. Noëlle looks very much the gamine, short-cut hair showing off a long neck and shell-like ears, a finely boned face with warm brown eyes, which fits perfectly atop her slender form. Long-limbed and graceful, she brings to mind a wading bird, a heron or egret perhaps, at rest then suddenly bursting into fluid flight. Right then, she appeared to be bursting with something else. She opened her mouth as if to speak, took a deep breath, then turned away, obviously working hard to stop herself from blowing up completely in front of everyone.

"I'll go by after service, talk to her, see what's going on," she said finally.

"Is that a good idea?" Jacques asked. "Let her sit. We can talk about it. Maybe tomorrow."

The rest of the help had fallen silent as well, the only sound knives and forks on plates, a bottle clinking on a wineglass.

At this point, they were just going into the busiest weeks of the busiest month of the entire three-month summer season. Anne was their longest-serving employee, someone who could turn her hand to anything and had: dishes, bar, kitchen prep, and waitressing. So not only were they left high and dry at the worst possible moment   the newspapers were full of ads from area restaurants looking for help — but the traitor was from the inner circle. It was therefore a more personal, and deeply wounding, betrayal, as Jacques and Noëlle had taken her in, really, and helped her to get back on her feet at a time when she had given up on her own family. She was also the only year-round employee, and they had worked out a special schedule for her so that she could still get some unemployment while working the off-season weekend meals for them. In essence, they had invested a lot in keeping her happy and financially solvent at their own risk.

After a few moments of quiet during which all eyes seemed to be focused on plates, or out the window, Jacques spoke. "It's always the ones you trust who walk out. And always about the middle of August! Remember my nephew, Noëlle? He came begging for work, on his knees practically. And I knew I shouldn't but he swore he would work the season—we needed a barman—until the very end. One of our customers is the head coach of the Gourdon professional rugby team, and my nephew was a rugby player, a good one. So he's tending bar and meets the coach and then, phweet!, he was out of here like that, in the middle of August, because he's got a contract to play rugby in the fall!"

The phone rang again, Émile, a retired Parisian banker of whom Jacques says, "Eating out in good restaurants is his single pleasure in life." He is also a very loyal customer who has been coming since the very beginning and who has also brought many, many others to their door.

"*Oui! Ça va bien.*" Everything's fine, Jacques said, at which point the whole table groaned in unison. With Anne gone, everyone was going to have to work a little harder, pack a few more tasks into their twelve-hour days. Noëlle would begin to cut down the number of reservations for the coming weeks, trying to keep lunch and dinner to no more than fifty or sixty covers each.

Jacques had a long and meandering conversation with Émile, the substance of which, Jacques told us after he hung up, was that Émile would like to change his reservation from Saturday to Sunday. "He is very particular. He'll call again tomorrow and say, 'Eh, Jacques, what good thing did you find for me at the market this morning' He knows I shop on Tuesdays and Fridays. Then he'll call on Wednesday to say, we'll be three at lunch and not four covers. And then he'll call again on Friday, because he knows I was at the market. He'll say, 'So, Jacques. Anything good for me this morning?' "

"Jacques makes things up. The lobster! The artichokes!" Fred said, laughing.

"Finally," Jacques continued. "He'll call again on Sunday to say, 'Jacques, we're running a bit late. We'll be there say, twelve-fifty and not twelve-forty-five.' A good guy, but he doesn't know what to do with his money!"

At seven-forty-five, with everyone back in his place, Noëlle's cheery, distinctive "*Bon soir!*" rang out as she greeted the evening's first customers, a couple notable for the fact that he, at a much younger age, had been a top fashion model in New York and Milan. Otherwise, he was memorable because he insisted every morsel that passed his lips must be raw, *cru*, although he had not bothered to call ahead to inform Jacques of this.

In the kitchen Fred was laying out the last of the garnishes at the plating station (where each course and its side dishes are posed on a warm plate before saucing and garnishing) at one end of the pass, the stainless-steel counter that runs ten feet from just inside the kitchen doorway straight back parallel to the windows and that separates the stoves and big center prep table from the rest of the kitchen. There was a tray of dill, basil, and chervil sprigs; a bowl of chopped chives; plates of paper-thin half moons of lemon, cucumber, and carrots cut into stars; buttons of cherry tomatoes. He had already laid out the prep table's *mise en place*, every utensil, spice, and required ingredient they would need within arm's reach. Along the center of the table from left to right was a plastic bucket of warm water sprouting six *pochons* (little ladles for saucing), a pair of tongs, and two long, thin spatulas to manipulate meat; four whisks for sauces; small bowls of salt, of butter in chunks, and of olive oil with a paintbrush; and a pepper mill. Two big wooden cutting boards nearly covered the center of the table, long knives for slicing meat to one side along with a sharpening steel.

Jacques was throwing together a raw scallop plate for the model, working quickly and not at all happy to have to stop his last-minute prep to do something that should have been ordered ahead. He covered the bottom of a wide, shallow soup plate with overlapping layers of finely sliced scallop. "Where's the *moulin à poivre?*" he yelled, looking for the pepper mill that Fred was already holding out to him. He ground black pepper over the snowy scallops. He then covered them with slivered *truffes d'été*, (dark brown white-streaked summer truffles not nearly as pungent as the winter black truffle) and dill with bright yellow cherry tomato slices ringing the plate.

Noëlle came in, saw him preparing a single plate, and took a deep breath. "Ah, Jacques?" she said sheepishly. "The raw plate is for two? I thought you knew . . ."

"*Putain de con!*" Goddammit-to-hell! He threw down his knife, ran to the walk-in for more scallops, started over again with another plate. "I'm telling you, Noëlle, this is the last time I'm doing this for him. He has to call ahead." Noëlle grimaced and rushed off. "Oh, that really pisses me off. It just pisses me off," he said. "Once is amusing, twice is enough, and the third time it makes me want to shit." Fred and the two *commis*, Bénoit and Hélène, hid their smiles.

Noëlle returned, almost whispering this time. "Uh, Jacques? What are you going to do for the main dish, they want to know?" Jacques gave her a dark look that said, How the hell am I supposed to know? and she backed off.

What he produced was a simple, quite beautiful, off-the-cuff variation of carpaccio, slicing raw *pavé de boeuf* razor-thin, dripping a few drops of sherry vinegar on top of the deep red fan of meat slices, then curls of summer truffles, yellow cherry tomato halves, extra virgin olive oil, parsley, then shaved Parmesan, the whole sprinkled with a bit of chive garnish. It was as if he couldn't help himself, even when pressed and inconvenienced, from offering up a thing of beauty, and this for people who hadn't even had the consideration to call ahead.

"Don't put any croutons on the soup, Sandrine!" Noëlle shrieked to the "raw" couple's waitress in passing. "Croutons are cooked!"

After spending several weeks watching the finely tuned opera that is a restaurant kitchen during service, I would understand why Jacques was so choleric. It's a full fifteen minutes—and right at the start of service when he has a hundred things to do—to feed just two customers. And he has to come up with a whole meal on the spot from whatever is on hand.

"*Deux couverts, ça marche! Trois couverts, ça marche!*" Working, two covers. Three covers working! Noëlle shouted, slapping down paper on the top of the pass. In the time it had taken to prepare the raw plates, the flow of orders changed from a trickle to a steady stream. By eight o'clock, Noëlle was dashing in and out every two minutes between *bon soirs*, calling the number of covers, or diners at a given table, and then setting the slips on the pass.

Jacques or Fred, both working *le piano*, the huge flat-top stove with four raised burners at its far edges, the salamander broiler for browning on one side and the steam convection oven on the other, took the slips, bawling out the actual orders so each person responsible for a particular dish could note it and start it working.

"*Un croq, une mousse, une ballottine, un gratin!*" Fred barked. "*Et ça marche cinq couverts!*" Noëlle shouted in passing. Jacques took the slip from her outstretched hand: "Three *ravios*, one *rouget*, one *croq*!" He paused, "AND THEN! Three *ballottines*, one *gratin*, and one *pavé*!" Whoever reads the slip then tucks it under the bottom lip of the salamander where those working the entrées can consult it.

In the time the orders had been read, Sandrine had jumped to the soups, clattering out the bowls on the narrow soup station against the front windows. She filled each with an inch and a half of cold tomato soup, sprinkled croutons and garnished with a wafting of chopped chives, then loaded up, one bowl resting on a napkin on her forearm, one in her left hand, one in her right.

Throughout the evening, the phone rang constantly, Noëlle taking reservations at the bar for the next days. "I'm sorry, Madame/Monsieur/Sir, but we are full for the night!" floated into the kitchen. "We had to refuse the mayor of Cahors, on his birthday, once," she told me later. "We were full up, I mean completely, and we just can't make an exception. When a mayor or somebody important comes in, you can bet they get treated like anyone else."

As the first main courses came off the stove, Fred stood by the back window, loading a single plate with all the sides for any given table, since they don't change with the main dish ordered. A plate for five covers—a table of five—will therefore have five sausage-stuffed tomatoes, five small rolled mounds of precooked linguine in a clinging Parmesan-cream sauce covered by a quarter of a peeled petal-shaped tomato, and five squash blossoms stuffed with a mousse of chicken breast, cream, and chanterelle mushrooms. Each plate of sides goes into the microwave for a few seconds just as the meats are finishing so that everything goes out hot.

By eight-fifteen, there were thirty-five covers working, meaning, at any one moment, generally about fifty separate items moving out of the kitchen. Of the entrées, one is a cold salad that Bénoit assembles, another small squares of goat cheese in very thin pastry called *feuille de brik* sautéed gently and served over a cold salad (Bénoit and Hélène), one lobster ravioli simmered on stove-top (Hélène), one scallop-stuffed squash blossom garnished with a whole crayfish and two tiny crayfish tails (Hélène), one melon half filled with a southwestern muscat called Rivesaltes.

Jacques works the stove and ovens, helped by Fred. They tackle the fish main dish, salmon and sautéed fennel in a *feuille de brik* robe browned on one side in a sauté pan and finished in the oven; the lobster gratin, which is assembled, baked briefly in oven, then browned under the salamander, the chicken breast rolled and stuffed with chanterelle cream sauce, which is simmered in water before serving; and two meats sautéed, duckling breast and the *pavé*, both of which are almost always served very rare and so must be watched.

Long before service even starts, each cook will have gone out into the bar to check the reservation book, not just to see the numbers coming for dinner or lunch, but to estimate how much of the particular dishes they prepare they'll need to prep for and what they can possibly do ahead of time to save even a few moments when service starts. For Bénoit and Hélène, on cold entrées and desserts, this might mean how much salad greens to clean, how many goat cheese squares to make up, how much dressing, and then how many orders of strawberries for the *gratin de fraises*, which sauces need to be brought to room temperature, even how many saucing ladles need to go into the utensil bucket at their station according to the number of that night's desserts.

For Jacques and Fred, it's a numbers game, too, with Fred finishing the last of the *mise en place* and plating station prep while Jacques cooks off all the meats that can be at least started even before the orders comes in. The idea behind cooking off meat is to save time: a duckling breast, even rare, takes three or four minutes of sauté, two of which can be done ahead. Jacques estimates that a certain minimum numbers of duckling, *pavé*, or saddle of lamb will be ordered and so starts them just before service begins, then puts them aside. When each is ordered, he can speed it to the table in a shorter time, thus freeing up stove space (he's got only four burners and the plaque, the stovetop between the burners, also heated, used for warming) and reducing by one the twenty other things he has to keep in his head at once.

And this is why it never pays to take a chef by surprise. Either he divides his attention and attempts to cook off the meats while preparing the special orders, or he puts off the meats and gets clobbered if by chance most of the customers show up at the same time. Which that night, of course, they did, nearly forty diners arriving within fifteen minutes of one another. His timing thrown off, Jacques would find himself not even an hour later squarely in the shit, *dans la merde*.

By eight-twenty, the second wave's orders poured in without a letup, Noëlle seeming to pop her head through the kitchen door every thirty seconds. "Four working! Six working! Two, two working!" Jacques was calling the orders like a circus barker: "*Alors! Cinq menus. Un melon, deux croquants, un raviole. Deux saumons! Ensuite!* Following that—four ballottines, *un pavé*! Benoit, help Hélène and take the *croquants*. And who stole my goddamned scissors?"

"*Entrées à la cinq!*" Fred called, signaling Jeanette, one of the two waitresses, that her plates were ready. "First courses for table five up!" "Coming!" she yelled in passing, arms full of soup bowls. "And you can do the entrées for table twelve, too." Across the counter, Jacques scooped soft butter from a small bowl with his fingers and slung it two feet through the air into a sauté pan, followed by a duckling breast into whose fatty side he had cut a crisscross pattern.

Twenty minutes later, with almost all the night's eighty customers seated and eating, the kitchen had descended into a kind of controlled chaos marvelous to behold. The heat coming off the stove, the shouted orders, abrupt questions, even shorter answers, the smoke and flames of beef, duckling, and fish being seared off, the personnel hurtling like linebackers from stove to walk-in to prep table and bearing hot hot pans, wielding sharp knives, all as every aroma from burned lobster sauce to sweat to simmering strawberries assaulted the nostrils. There is, all at once, a bit of Dante, a dollop of Marx Brothers, and even a soupçon of ballet in this, and all before the plates reach the table.

Hélène, working her first real kitchen job after finishing her cooking degree, was clearly in the shit, way behind, her pale skin flushed and forehead sweaty, a few strands of strawberry-blond hair come loose and sticking to her cheek. She ran from the stove with two tiny crayfish tails on a spoon, dropped them. She was bending down to rescue them when Jacques kicked them under the table. "Just get two more! You don't have time." She did, deploying them on a plate next to a scallop-mousse stuffed squash blossom, then swiveled to grab a too-hot sauté pan with goat cheese packets browning. She cursed, reaching for a towel.

Next, she took a huge sauté pan and painted it with olive oil, then dashed to the fridge, emerging with a sheet pan full of red mullet fillets. She started putting them in the pan, ten fillets in all for five orders, when Jacques peered over her shoulder at the sizzling fish. "When they're big like that, the fillets, don't put so many in the pan. They won't cook evenly, okay?"

Jeanette dashed past, shaking a pepper grinder at ear level to see that it still had peppercorns in it.

A butane kitchen torch ignited with a loud pop as Jacques browned a lobster gratin the quick way. An order of ballottine, a salmon, and two nicely toasted gratins waited on the pass for him to finish, the last of the table's order. "Which next Jacques?" Frédéric asked, holding two order slips in his hand. "The two deuces or the five?" "Do the five!" he called from the stove, poking two *pavés* and three duckling breasts that sputtered in another large sauté pan with his finger to test their doneness. "You can always cook them more, but you can never cook them less!" he said, seeing me watching.

The center prep table, meanwhile, held two large cutting boards on which a pile of *pavés* and one of *canettes*, all of which Jacques had cooked to within two minutes of rare, waited to be finished when ordered. Next to this, fry pans filled with mullet fillets, plates with frozen raviolis, more sauté pans with goat cheese squares, a precarious pile of sausagelike ballottines — every available surface was covered with pans and trays and cutting boards of entrées and main dishes on their way to the crowded stovetop or into the oven or under the salamander or onto a waiting plate. Down one side of the table, a line of order slips grew, each one placed there after the entrée goes out and the main course starts working. When Jacques plates the entrées, the slips then migrate to the back of the room to flutter over Bénoit's workspace, the only thing left to do on them being the desserts for which he and Hélène are responsible.

This is the time when the waitresses begin to "gallop," as Jacques says — begin to get crushed, especially that night as Anne had quit and they were consequently one short going into the crush leading up to August 15, a French national holiday. Jeanette, Sandrine, and Noëlle streaked by in blurs, bearing bread baskets and soups, carafes of wine and water, the ice buckets for white wine with their refrigerated inserts and armfuls of entrées and the first main dishes as they came up.

"Monday is hard," Noelle remarked during a one-minute lull. "Especially after a very busy weekend like this last one, because it's the day before marketing. So that means there's less choice than the customer thinks. It may be on the menu, but we have only so many orders of it. At eight o'clock, I know we still have twenty-five more people coming in later, so I want to save some of the really special dishes for them. Jacques puts everything—all of the good stuff—into one dish sometimes," she said disapprovingly. "Look at the ballottines! They're stuffed with chanterelles and THEN finished with a truffle sauce. Chanterelles and truffles—of course everybody wants it, so I have to steer them a little. 'And if you'd prefer something light tonight, we have a marvelous croustillant of salmon on a bed of soft fennel!' So, tonight, that big table of twelve, three of them ended up taking the salmon instead of the ballottines."

"*À la cinq, mes gars!*" Jacques called out, time to plate the main dishes of the first big table of ten. Jacques and Frédéric stayed behind the pass, Jacques to shuttle the meats from pans and oven, while Fred set out the sides and Bénoit and Hélène raced to garnish from the other side with the four and sometimes five elements that make up the decoration of every plate. Fred pulled down one sauce pot after another to do the final saucing, then Jacques finished with the final sprinkle or sprig of garnish depending.

"Jeanette!" Jacques called. "*Tu peux commence!*"

Jeanette blitzed in, muttering to him to sod off. She is English and, in her second year in France, speaks the language moderately well. He was mocking her French from the early days, doubly wrong in that not only should she address him in the more formal *vous* but she tends to use every verb in the present tense, saying, "You can are starting." She takes the ribbing and gives as good as she gets, as a result of which Jacques now has a good supply of Anglo-Saxon curses at his disposal. As plates were finished in threes and fours, Sandrine and Jeanette swooped in, grabbing them up and bearing them off to the waiting table.

As quickly as the crowd had gathered, it dispersed, everyone back to his tasks. It was only nine o'clock, and there was still a lot of work to do. Outside, ten people tucked into ten first courses, all hot, all beautiful, and the time elapsed between delivery of the first and the last just enough for the preparatory adjusting of napkins and palate-clearing sips of wine.

A minute later, Jacques turned from the stove. "Hélène," he asked, "Why are you starting the raviolis for the twelve? They don't even have their soup yet!" Always intense, her face clouded, and she froze. "It's okay, just put them aside. Why don't you do the soups for them instead?"

Jeanette, her face running sweat and her white shirt blotched in the small of her back, saw Hélène and Bénoit pouring soup and garnishing and smiled at them. "That's for me, Bénoit. You are too kind."

At nine-ten Jacques, interrupted once too often by people wanting to know if there's room, if they're open, if they can make a reservation for next week, unplugged the kitchen phone. Each time it rang, he either had to summon Noëlle or leave the stove to go out to the bar in search of the reservation book. Shaking his head, he opened the oven door to check some gratins, and a wave of heat rolled out.

With the diners all seated out in the courtyard facing the wide tall windows of the kitchen, I noticed a curious thing. The chefs, in their one minute of downtime every half hour or so, always stopped to gaze out into the court, watching their labors being consumed. Of course, they also kept up a running commentary. "Did you see that one in the black, you know..." "With the breasts, you mean?" "Bénoit was looking at the one next to her, right?" "Jeez, Bénoit, she must be all of fifteen!" Bénoit blushed. "Was not!" "Were too!" Hélène studiously ignored them while Sandrine ragged them for screwing off when there was work to be done. It's at moments like these that you realize the quiet, serious, hardworking *commis*, these chefs of all work, are not much more than teenagers, the world of the kitchen their only education from the age of fifteen.

The diners in the courtyard, meanwhile, watched their own show, heads up from their plates to catch the action in the kitchen, perhaps listening in on the mêlée taking place there. "*Oui, oui. Ça fait un spectacle!* It's like the theater!" Jacques said.

At a little after nine o'clock, the first cheeses began to go out, a signal to Bénoit that he'd better get his dessert plates set up since that course comes last. On the right side of his long stainless-steel table, he was still putting together goat cheese salads, while on the left he began to lay out clear glass dessert plates. Ten minutes later, he'd gotten eighteen prepped, each with a dusting of powdered sugar, a single slice each of starfruit and kiwi, a tiny bunch of carmine gooseberries, and a single, perfect raspberry. Then he stacked a second layer in an overlapping pyramid, took out the sugar, and began again until twenty-eight plates were prepped and ready to be filled with any one of the four desserts served like this.

"Let's do the *douze*!" Jacques called out. Again, everyone came over, this time to plate the first courses of the second big table, twelve diners.

By nine-thirty, there was one table of six that had not yet arrived, and the crew was getting restless. There are certain un-written rules to eating out a good restaurants in France, and this is one of them: Don't arrive late, at the close of service, and expect to be fed. Why? "If they come in at two-thirty for lunch or ten for dinner [the close of service], our day is not thirteen hours long but fourteen or fifteen hours. That's just too long."

A few minutes later, crisis. Surveying the meats just finishing for the big table, Jacques looked puzzled. Checked the slips. "Shit. We need eight ballottines, not seven. *Putain!* Goddammit, Jeanette! You said *la table dix, dix, non*?" Table ten, she said ten, right?

"Houston? We have a problem!" Fred called out, his single phrase of English, looking for the waitresses. "*Six, c'était la table six?*" Fred asked Sandrine, who was the only one not busy. Table six, that was? Since *six* and *dix* sound so similar in French, Jacques was wondering if Jeanette had mixed them up. He looked over and saw that Fred had done the correct number of sides. He lined all of the ballottines up on the cutting board; seven chicken breasts stuffed with chanterelle farce and rolled into sausages, then sliced into delicate rounds. He began plucking one slice from each of the seven to assemble an eighth serving, fanning the slices, nodding. It's okay. After plating, he was extra generous with the sauce, which has bits of black truffle dotting the pale chicken like tiny currants.

Hélène, Bénoit, Jacques, and Fred, an assembly line of garnishers, gathered around the thirteen plates on the pass, most of which were for a single table.

"Everyone!" Jacques called to the waitresses, saucing the last of the ballottines. "Give a hand for the big table."

Sandrine had pink in her cheeks, strands of fine brunette hair had come loose from her bun, and she blew them out of her eyes as she covered her left forearm with a towel and laid three plates on it.

The phone rang. Noëlle must have put it back on the hook. Jacques picked it up, frowned. It was a friend calling for a reservation. "Hey, you know," he said with exasperation. "It's really not the time to call me." And hung up.

With the pass once again empty, Jacques spoke his mind to the waitresses, and to the crew in general, about keeping the orders straight. "You can't screw around when we're busy. We have to remember what we're doing here, okay?" Heads were hung, but then the storm passed. Jacques is like that, and his kitchen is like that. You screw up, you get yelled at, you fix it, you move on.

At nine-forty the last table, for whom they'd all been waiting, finally arrived. "Jerks!" Noëlle said, handing over the slip. Jacques called the order and faces fell. And just when they thought they might have been in for an early night. Now they would be at the stoves for another hour at least, and one waitress would not finish until midnight or even one A.M. if the latecomers lingered for coffee and *digestifs*, which they always do.

Bénoit's torch cracked as he browned crème brûlées, then he was rushing to the salamander to rescue a handful of apricot *clafoutis*, the light custard covering the fresh fruit seconds away from burning. Fred helped as he tended the meats.

Finished third courses were meanwhile going steadily out the door. Sandrine came in. *"Viandes pour la six!"* Meats for table six! Fred yelled, even though she was right there.

"That's right, table six!" she repeated, looking at Jacques.

"That's right, table six!" Jeanette repeated emphatically, also looking at Jacques, underlining the point that it was he who had earlier screwed up the number of ballottines. This was just another shot in the long war between front of the house and back of the house, each fighting over who would attempt to control the pace of the service even though it is ultimately the customers who determine how fast or slow things move.

"So I made a mistake," Jacques said, shrugging his shoulders, a sheepish smile crossing his fatigued features. "It happens, it happens!" He gathered himself, turned back to the stove. "Let's go disco!" he said in English, sort of.

"Et, voilà!" Jacques said ten minutes later, unslinging the pair of long tongs he keeps stuck into his apron strings and chucking them into the sink. (Fred keeps a thin-bladed metal spatula stuck in his.) He wiped the sweat from his brow, reached for his cigarettes. "You see how a chef can drop dead in a kitchen like this at age thirty-eight?" he asked, referring to a recent newspaper article about just that. "And there are restaurants that work like this, the same size, but year-round!"

Noëlle came in, stopping him in the doorway of the kitchen, huffing with outrage. "They're drinking Coca-Cola with their lobster! And they're French! *C'est un sacrilège!*"

Jacques shrugged, just happy to be finished. The night had turned cooler, and he was on his way around the back of the restaurant to go get Nougat, his dog, from the cave, or wine cellar. From inside the kitchen, Fred, Bénoit, and Hélène could see him out the back window, rough-housing with the dog, then lying on his back on the grass, looking up at the sky and smoking.

In the kitchen, Noëlle was still ranting. "The idiots at six, they asked for a rare steak with béarnaise sauce, but sauce on the side." She drew herself up, her most formal and just a bit intimidating. "I told them, 'We don't have steak, but *pavé*, and here we only make our own sauces, and a béarnaise was not one of them, but that we would be glad to put the sauce that comes with the steak on the side.' "

After a moderate start to the evening, Bénoit was really earning his keep, bent almost double over the clear glass dessert plates, fingers deftly dealing garnishes, saucing, clouds of powdered sugar and minor explosions erupting from his station as he powered through the fifty-odd desserts of the evening. They tend to come clumped together at the end, too, those who arrived early lingering in the cool summer evening, savoring the breeze, not hurrying their meals.

Jeanette was also hustling, taking five *cabécous* at a time, each disk of goat cheese garnished with a single fresh bay laurel leaf from the bush in the front courtyard.

Waitresses do run, as do chefs. Jacques measures the pace by describing the service as either a promenade, a trot, or a gallop. "They're not feet by the end of the summer," Jeanette said one day. "They're bloody stumps!"

By ten o'clock, Bénoit was down to eight plates, Hélène torching crème brûlées beside him. Fred and Jacques, since back from his smoke, were torturing Jeanette, repeatedly blowing out the single birthday candle on the *bavarois* for the fifty-year-old at table seven. "Bugger off!" she said finally, elbowing them aside.

A few minutes later, Bénoit, Jacques, and Fred stood in a line at the window, scoping out the women in the courtyard. You could almost tell the country of origin by the clothes, the French women going in for low-cut, slinky, sheer, sleek, minimal dresses and skirts and tops, the English ladies looking more like they last bought clothes twenty years ago, the Dutch most casual of all, the German grandes dames looking well put together, rich.

"The dark-haired one? The dark-haired one. Look at that dress!" Fred said. "A real cannon!" Jacques replied, catching a withering look from Noëlle, coming in for more desserts.

By ten-thirty, everyone was trashed, energy flagging. They had started the day at eight-thirty, did sixty-plus meals at lunch to finish at three, and eighty that night. The crew, momentarily revived by tall glasses of cold water with mint syrup, sent Jeanette out to bother the one guy at the last table who hadn't finished his first course so they could finish cooking the table's main dishes and then start cleaning up.

Jeanette came into the kitchen after clearing the "raw" table. She had an empty *gratin* of strawberries plate in one hand, which looked like it had been licked clean, and an empty breadbasket and a full ashtray in the other. "Must be a whole pack of Marlboros in there," she said dryly. "And they didn't have any problem with the bread and *gratin*, either!"

The summer season is, for Jacques and Noëlle, both blessing and curse. On the one hand, the restaurant turns over something like six hundred meals a week for the peak period from June through the end of August, slightly fewer in May and September. In October, November, March, and April (they close from December through February) they operate with a skeleton staff for their three-day week, sometimes just the two of them with one *commis*, or calling in an extra waitress as needed. When the fall weather forces the diners inside, the maximum number of seats that fit in the dining room is only thirty-five, still no cakewalk for two chefs preparing a five-course prix fixe menu. Jacques' mother often drives up from Toulouse to give a hand with prep and the dishes for the five weekend meals they serve in those out-of-season months.

Although their success has allowed them to hire more hands in the peak period, the hours required of both Jacques and Noëlle, and to a lesser extent the rest of the staff, are prodigious. The average workweek is seventy-two hours, five and half days a week from eight or nine in the morning through midnight with a three-hour break in the afternoon. Jacques and Noëlle put in even more time as their day off is consumed in maintenance, banking, and all the other things that can't be done while the restaurant is running.

On top of that, twice a week Jacques is up at four or five in the morning to drive to market in Toulouse, an hour and a half away.

Such a superhuman effort takes its toll, and not only on the body. "Some nights, especially by the end of the summer, I can barely make it up the stairs to bed, the cramps in my legs are so bad." Jacques admitted. "And once he's in bed—PAAF! He's out," added Noëlle. "It's like sleeping with a log in the center of the bed! He always loses a couple of kilos, too."

The day after my restaurant initiation, we were hurtling down the A20 at four in the morning in Jacques' white, refrigerated Peugeot van. Marketing days in Toulouse always begin like this, a brutal rush from before dawn to get back in time for morning prep at nine-thirty. And then an equally grueling prep, turning those ingredients into the dishes that customers will be ordering in just a few short hours. And then lunch service in front of the stove until two-thirty, when he can crawl upstairs and fall into bed for a quick nap before dinner service.

"I don't cook anything new in the summer," he was saying. "There are just too many people, and we are just too busy for me to take the time, to even *have* the time to reflect on what I'm doing, never mind coming up with something new."

I looked over and saw his haggard profile in the intermittent light of the streetlamps. The blue eyes seemed more deep set, surrounded as they were by dark circles of fatigue. The crow's-feet at the corners were more pronounced, the silvery glints in his dark hair more numerous. His skin looked gray, drawn tightly over his cheekbones, his shoulders a little bowed. He tapped the wheel restlessly as he drove, the muscles in his thick forearms rippling, then lit up a cigarette. He had managed only a couple of hours' sleep after service, rehashing the Anne situation with Noëlle and wondering how they were going to make it through the next weeks when they should have been sleeping. With seven salaries to pay before they touch a franc themselves, it's a large burden to bear.

"I don't know." He sighed. "Nine people on staff, it's something of an experiment. Yes, we can serve eighty people lunch and dinner, but we'll have to see at the end if maybe having fewer staff and serving fewer people makes more sense. And the new ones, they're young. They break a few more dishes. *Ils sont fou-fou*—I have to concentrate their minds on their jobs. Bénoit, last year he hardly opened his mouth, very serious, didn't say a word to anyone. Now he won't shut up, and he's got a voice which carries. The dining room is just next door and I have to tell him, '*Doucement!*'

"I sometimes have the impression that the summer crowd would eat anything I put in front of them, at any price." Jacques said this with despair. "So why should I bust my ass for them?" He fell silent, watching the landscape pass in a blur. "Because," he said a few moments later, softly, almost to himself, "that's what makes it a good restaurant."

THREE

MONSIEUR LE MAIRE

I once asked Raymond Laval, a former farmer and the mayor of Les Arques, what word he had used for "weekend" before the Anglicism "le weekend" arrived in the French language. He cocked his head back, thought for a moment. "Sunday!" he said cheerfully. "We always had to work on Saturday, and so we didn't have any weekend."

As September rolled around, the atmosphere of the Lot changed almost overnight. In short, everybody left, and it began to rain. In Cazals, the Sunday market shrank to half its former size, the café was empty, and there were no longer any kids for Lily to play and swim with at the recreation area just down the main street. No more the fêtes and festivals and fireworks every weekend in this village and that to enliven every week and draw the hordes. Restaurants and campgrounds closed, and the whole region, it seemed to us, began to turn in on itself as what felt like entire weeks passed in endless cold, soaking rains. In Les Arques, half the houses closed their shutters, the extended families that had gathered for summer dispersing until the next long holiday, which didn't come until the end of October, and those owned by foreigners abandoned completely until the next summer. Slowly, the true nature of the Lot, at least ten months of the year, was becoming apparent. It was seeming a rather desolate, dreary place in which we had few acquaintances and even fewer friends.

The exception was the Laval clan, the mayor and his extended family, who adopted us almost from the start. Because I had had a certain amount of contact with minor French functionaries over the years, my initial hopes were not very high. I expected a perfunctory, rather distant stuffed shirt going through the motions of helping me in the name of upholding his office and showcasing his village, and not much else. I would end up almost embarrassed by how wrong I had been.

White-haired and elfin-faced, with the energy and appearance of a far younger man, Raymond Laval is about as far from the overly formal, bourgeois, fustian stereotype of French officialdom as you can get. He is almost five feet four inches, whipcord thin, with sinewy forearms and slightly bowed legs, and his face has been burnished a fine shade of red by constant exposure to the sun, which has also left its traces in the deep crow's-feet at the corners of his dark eyes. He looks infinitely more comfortable atop his much battered 1975 McCormick 323 tractor in a faded shirt and tattered straw hat, off to tend his walnut trees or to work his vines, than in suit and tie officiating at a village ceremony.

He was our closest neighbor, his house at the hamlet of La Mouline, five houses and one mill just down the road from us. His house and the long barn in front of it were originally part of a wine depot, a collecting point for the iron-banded barrels of chestnut and oak that would come down from the surrounding hills in wagons drawn by horse and mule around harvest time every year. The drive leads right up to the great arched doors of a barrel-vaulted cave, a cool, dark cellar built into the back of the hill behind the house and taking up its whole lower story. Over the door the date 1761 has been chiseled into the stone; over the fireplace lintel, 1767.

Though varying in the details, the broad outline of his life would apply to most of our neighbors. "I was the only son of parents who were also only children," he told me one day. "I was born in this house. We had *un petit élevage*, a smallholding of cows, sheep, chickens. Some forage crops. We had twelve cows to sell the milk. We grew tobacco from just after the war until the 1970s. We all lived there, on the farm, at first myself with my parents, then my wife, and then with our two sons. In 1958, I left the farm to go to work for the Public Works Department, maintaining the roads and bridges of Cazals and Salviac. I did that until 1983, when I retired. Many families did as I did, leaving the farm when properties became too small to support so many people, the land too poor to offer more of a harvest."

Raymond talks like that, very simply, directly. He is a supremely quiet man, calm in his bearing. In 2001, at age seventy-six, he retired from his third career; after twenty-eight years as mayor of Les Arques, he finally passed on the tricolor sash in March to Patrick Cantagrel, the son of one of Raymond's childhood friends who died in the fall.

Monsieur le Maire, as he was first formally introduced to us (and how my daughter addresses him to this day), was the first person we met in Les Arques, not fifteen minutes after arriving, taking us to the house we had rented from Gérard, his son.

As he showed us around the small farmhouse and turned on lights and explained this and that, I took in not a word. Rather, I listened, for Raymond speaks both softly and in a French heavily accented by the local patois, which was his only language until he went to school at age seven. This gives his words a lovely Mediterranean rustle about the front of the mouth, stripping out the nasals and instead substituting hard consonant endings. The simple phrase "very good," *très bien*, emerges from his lips sounding more like Catalan or Spanish, the initial "tr" formed with the tongue against the teeth rather than in the middle to the back of throat, while the "en" of *bien*, in standard Parisian a foggy nasal, here is both clipped and hard, with the "n" enunciated. Spoken rapidly, this patois has an impenetrable but quite agreeable rollick to it.

Throughout the early days of summer, we would see Raymond puttering up the road on his tractor, Suzanne, his wife, following behind in the rattletrap LeCar, on their way in the cool of the evening to work their large garden across the field from our terrace. They had long rows of lettuces, tomatoes, onions, leeks, beets, turnips, radishes, green beans, cabbage, and potatoes. And Suzanne had put in a row of cutting flowers, too, their colors very bright among the green of so many vegetable plants.

We had had a few short conversations in the first weeks, and I had spent an awkward afternoon with Raymond in his office, asking about the recent history of the village. It was awkward for many reasons. The French are reticent in general, and the country French are doubly so. Add to this that I was not only an outsider but a foreigner writing a book whose nature he didn't quite understand, and I'm surprised he talked to me at all.

One evening, unable to bear watching the two of them laboring any longer as we lolled on our shaded terrace with cold drinks, we went to help. Raymond had just turned over the earth under two long rows of early-ripening potatoes, and my wife and I took up places beside them, plunging our hands into the warm soil, digging out the lightly buried harvest. They looked a little surprised at first, but Lily immediately made a game of collecting the very smallest, some the size of marbles. We talked as we filled the buckets, as the buckets filled burlap sacks, as the sacks gradually filled the box on the back of the tractor and the light died away. The next morning a basket of the new potatoes, washed and dried and gleaming in their red and golden jackets, appeared on our terrace wall.

What did we talk of? Nothing important or deep, but of where we came from and how we grew up, of brothers and sisters and mothers and fathers, of the beauty of their land and the contrast with our own rocky Maine coast. Two families, one young and one older, finding common ground in the dirt at our feet and the land all around.

A few days later we found ourselves one afternoon sitting

around the scarred wooden table on their porch, snapping the stems off a couple of pounds of just-picked green beans, some of which ended up on our dinner table for the next several nights Suzanne asked almost hesitantly how we were settling in, what we were finding to do on those hot summer days, while Raymond looked on, smiling and offering a few words here and there, pouring a glass of the local red wine cool from the cellar, as they like to drink it here.

Suzanne is just as diminutive as her husband, a square face framed in puffs of fine, white hair with a strong nose and chin, a woman who smiles frequently and looks you in the eye when she talks, who lays her hand on your forearm and clasps you firmly by the shoulder when she gives the traditional kisses of greeting. Patient and soft-voiced, she seduced our daughter immediately, cajoling her to speak French.

Earlier, she had taken Lily by the hand, and off they went, Suzanne filling her eyes with the surprises of the farmyard, the chickens on their eggs, Toby the retired sheepdog. There was a delicate moment in front of the rabbit hutches. Lily went right up to the cages, giggling as a rabbit snuffled her outstretched fingers. "They're so cute!" she said in English, at exactly the same moment Suzanne said, in French, "They're so delicious!" My wife and I exchanged glances and hid our smiles. As Lily wandered off to collect the eggs, hand in hand with Suzanne, I realized that she had made her first friend, and, what's more, found a substitute *mami*, a grandmother, to begin to melt away the strangeness of this place.

As the pleasant afternoon drew to a close, we did what we would have done at home, newly arrived and finding ourselves in good company. We invited them to dinner a few days later.

They appeared, carefully but not formally dressed, bearing gifts, of course, a stuffed animal for Lily and a bottle of wine, its cork sealed over with red wax, no label except a paper tag around its neck saying 1996, the fruit of Suzanne's brother's vines.

We opened the whole duck foie gras that I had bought from a farmer, Luc Borie, in nearby Catus. They both tasted it, then

Suzanne said it was good, but not as good as goose foie gras or even her own. "This is a small foie gras, duck, and with goose the texture is finer, more solid. You come to my house sometime, and I'll have some real foie gras for you."

"Don't tell me you prepare your own foie gras?" I asked her. "*Si, si!*" she said, as if it was out of the question that she would NOT make her own foie gras.

It was a little formal at first, you sit here, no, here, let me take your coat. They waited to eat until I took a bite of foie gras, touched glass to lips only after I had taken a sip of wine. It was quite clear that they had had no idea what to expect. But they soon loosened up, with Suzanne embarking on a long motherly retelling of her two sons' childhood and how they had arrived at their present circumstances. Gérard is still single at almost fifty, a designer and site manager for construction and renovation projects, mostly the older houses being done over as summer houses for northerners or foreigners. Alain, younger by a few years, runs a driving school with his wife, the lovely practical Michelle, who also puts in one day a week at each of three town halls where she is the secretary. They have two boys, Yann, twelve, and Melvin, seven. Both grown sons live less than a mile away from their parents. The whole time she was talking, Raymond was looking at her with a sort of wondrous affection, alternately nodding his head slowly at the high points and shaking it at the low.

What had brought us to France? Suzanne asked, which was really a question for me, about what my book was about, a curiosity they were too polite to reveal head on. We explained our long association with the country and its food, falling in love in Paris as students, eloping to a small village in the Haute Garonne six years later.

They had met at a dance in Suzanne's village, Gindou, five miles over the ridge. "There were no cars, certainly not for the young people. Some had bicycles," Raymond said. "And so the girls you knew were those who lived at most a couple hours' walk away, in the surrounding villages. You got up on Sunday, went to church, and then you walked to whatever village was having a fête that night, and ate a good meal, and danced!"

"And then you walked home," Suzanne finished, "because you had to get up to work the next day. Not that there weren't a few who ended up sleeping in the ditch by the side of the road." She looked at me, then to Raymond, saying, "But they were young men!"

Raymond was born, grew up, and raised his family in the house down the road. He was twenty-five years old when they married, and brought his wife home to live with his parents, which was what you did then. "We didn't have any money to buy a house, our own land," he said. "No one did. And that was always the problem. Not enough land when the next generation came along, and the farms getting smaller and smaller." They had not even had enough money to share a house with another family, which was also common, a door and a chimney at either end and a wall in the middle dividing the space in half.

"And it was not always a good thing, what necessity forced on you, so many people in a small house." Suzanne added meaningfully, "You don't have any intimacy." That was the closest I ever heard her to expressing regret for what had been a very difficult first decade of marriage, with little money, two babies, and in-laws to feed and care for. Later, when I learned she had carried every liter of water the household consumed by hand from the well at the bottom of the hill, I began to understand just how hard those years must have been.

I asked about the area during World War II, when it was part of Vichy, until all of France was occupied in 1942. Raymond was just finishing his education at the village school when the war began. "Oh, I remember we all had to go to the only house in the village that had a radio, it belonged to an old woman, and we listened to Maréchal Pétain. And then we sang a song to praise him!" This last he added with a bittersweet look, as if to say, Can you believe that?

In 1940, France capitulated, Vichy ruled this part of the country, and people went back to work the land, for otherwise they would have starved. He joined the resistance, ready to flee into the woods whenever the Germans came through as they did more frequently the last two years of the war. "They were looking for us, for all the young men by that point," he said. "Vichy was over and all of France occupied. We didn't want to be sent to Germany to work. We were already hearing that they were bombing the factories, too."

By 1944, they told us, the retreating Germans, realizing that they had lost the war and more frequently harassed by the resistance, were desperate, nervous, and on a hair trigger. In the village of Frayssinet-le-Gélat, this led, on the afternoon of May 21, to what the locals still call La Barbarie Allemande, a massacre of fifteen villagers that may have started over something as simple as an insult shouted out an open window. Months later, Madame Lafarge, who lives over the hill from us, would narrate the whole harrowing scene to my wife, people being pulled out of their houses, others fleeing into the woods. The villagers were herded into the church, then the victims picked at random, lined up, and shot while she watched from a hiding place outside the village. And then the Germans made the villagers bury their own dead. "And the ones who did survive, from the village," Raymond told us, "it was because they were up here, playing a game of basketball against the Les Arques team! We all heard the firing, and then people came running in, telling their friends not to go home." Suzanne was shaking her head back and forth, back and forth. "It was a terrible time," she said. "A terrible time."

Later in the evening, I found Suzanne looking around the more or less open-plan first floor where we were eating. "It's just so strange to see this house like this," she said, and proceeded to tell us the story of this place we had rented that once belonged to them and that they had passed on to their son. A childless farming couple once lived here. When he died and she was already aged, the widow asked Raymond and Suzanne to take care of her, to feed her, until her death, upon which they would inherit the property, a country form of Social Security.

What is now our kitchen had been a *saboterie*, where a local man made the wooden clogs everyone wore to work in the fields, they told us. The covered terrace was *un four à pain*, a bread oven shared by all the surrounding houses. In front of the fireplace there had been a trapdoor cut into the floor that fed directly into two great cement wine fermenting vats that still take up half the cellar.

After they had left that night, as I was falling asleep, I tried to imagine what it must have been like to live here then. The upper half story with its chestnut beams arching gracefully over our bed, this would have been the home of roosting pigeons, with gaps left in the walls so they could come and go and roosts to encourage them to stay, the floor covered in the guano so prized for fertilizing the grapes. In the single large room downstairs, the family would have slept, eaten, played, worked. With the house all closed up against the cold, the air would have been heavy with smoke from the fireplace, the reek of the oxen and milk cows filtering up through the floorboards together with the powerful mustiness of the grapes merrily fermenting in their vats.

It didn't take long for me to realize that Suzanne and Raymond, while absolutely representative in their lifestyle of the people who lived here, were very different in some ways. I heard an Englishman who had lived here for decades joke that, after a nodding acquaintance with his near neighbor of twenty years or so, the other man had finally began using his family name. "I expect," he continued, "that in another twenty he might come to call me George. Of course, I could be dead by then." In my own experience, I would say, *"Bonjour!"* and introduce myself as the American who had rented La Forge. They would say, *"Bonjour."* And I would wait . . . and wait. There was no malice in it; they just didn't have the habit. And they didn't have the habit of it, Suzanne told us, "because we already know everyone we need to know." If they didn't show a deep desire to get to know you, it was because you were a stranger or a vacationer or a summer resident with whom they would have little contact anyway.

But Suzanne and Raymond were not like that. They were curious. At first, I was hesitant to interrupt Raymond at his work until I realized that he liked it, liked that I felt comfortable enough to trudge through the muddy fields in my rubber boots to chew the fat with him while he leaned on his hoe and took a break from the dozens of tasks that consumed his day. I had so many questions about what I saw on the land, and here was a man with most of the answers. And the easiest way to get him talking was to be there, on the land, to take a walk together, sometimes to lend a usually awkward hand in what he was doing.

The landscape of Les Arques, indeed much of the Bouriane, as this rocky, rugged central part of the Lot is known, is a puzzling one to anyone who has spent time in a rural, agricultural place. First there are the farmhouses and outbuildings, so many of them abandoned, their roofs falling in, stone walls crumbling, drives grown up with brush. The fields around them, if maintained at all, have gone to hay, a frequent sight in late summer the great rolls of moss green tall as a man, drying before being wrapped in plastic and hauled off to the barn. Mostly, however, the eye catches on the ragged forests, of chestnut on the richer, wetter land, scraggly oak on the drier crowns of hills and where there is little soil. None of these trees, in vast swathes, are more than thirty or forty years old, and, in stark contrast to the orderly plantations of poplars in the riverbeds, all grow cheek-by-jowl, chockablock, the understory choked with brambles and ivy and brightened by the occasional patch of deeply green pungent wild boxwood.

In the middle of these young woods I often came across farm equipment just left there, hay balers and discs and plows and harrows, some of it from the era of forged and cast iron, spidery spokes of huge wheels, their rims made of bent iron bands, the intricate gears from an era where it was the turning of the wheels themselves that provided the energy to run the equipment. But much more of it looked more modern, if not of our time, obviously designed to be run off the power train of a tractor.

The Lot is a walker's paradise, for there are farm roads leading to the most distant, hidden corners, tracks with tall grass growing down the middle from the passage of tractors and wagons, and wide paths still visible through the middle of the young forests. Most lead nowhere, only to another abandoned field or pasture or farm. Bushwhacking on the drier ridges I came across dozens of *cazelles* and *garriottes*, round or oval shepherd's huts built with only mud for mortar, rude walls four and five feet thick at the base, each stone interlocking with its neighbors to support the whole, their roofs built up of successive layers of flatter stones, each layer overlapping the one below by only an inch or two. They were marvels of simplicity, and necessity, more like huge piles of stone with a doorway leading to a hollow the size of a man inside. Most of them were falling down.

I also would find the occasional stone-lined pit, half filled with fetid water if it had rained recently. It had served for something, but what, I wondered, lost out here in what was now scrub and brush. Halfway up a hill, I would realize the land was undulating gently, then set my eyes, half closed, along the ribbon of land from ridge to valley, seeing the bramble-choked and eroded terraces emerge. Out in the woods or grown-over fields, I often had the impression that, at a certain moment, most everyone had just up and left, leaving everything behind until it had become just another part of the landscape.

The Lot has always had a tenuous relationship with prosperity, and never more so than during the last hundred years. Dominated by the hilly limestone uplands called *les causses*, the only truly arable land is in the narrow valleys or on the very rare plateaus with sufficient topsoil. Grapevines are one of the few useful plants that thrive in the less than fertile, chalky soil and very dry summers here, and thus much of the village's modern history has been determined by the grape.

As I was driving along the roads for the first time, what struck me most was how very small and irregular the fields are, and how they are tucked into every available flat, or even relatively flat, space. Marie-Josée, Yvonne Trégoux's daughter, cultivates three fields at Auricoste, two of about an acre size, the other half that, alternating between corn, barley, or rye, and every third sowing a soil-building crop like clover or grass to let the land recover.

On the drive up from the mayor's house to the next village to the east, Montgesty, there are fields even smaller, and most more bizarrely shaped. The most appealing to me follows a stretch of climbing road with two tight S-curves. The field snakes along between the road and a steeply rising rockface on the far side, narrowing at some points to less than twenty feet across, perhaps three times that at its widest, for more than half a kilometer. And each bit of field seems to be used for a variety of crops—corn, barley, oats, sunflowers, rye, rapeseed—which run right up to the edge of the road, with only a ditch separating the two. Clearly, this is not large-scale commodity-driven monoculture, and also clearly, every bit of land is precious. Why, I wondered, had so much of it been so recently abandoned?

When the farmers brought in the harvest in the fall and turned over the fields to sow them with overwintering crops, I had a few answers. The odd corners and modest size are due to the fact that most fields are filled with rocks! I would come to realize very quickly when someone was plowing anywhere around us, even kilometers away, from the crash and clank of the steel blades hitting limestone. From pebbles to fist-sized wrack to flat chunks of a hundred pounds, the harvest of every plowing is another crop of stones gleaming whitely amid the iron-tinged red of the soil. Some of the fields, especially after a rain, appear to have been freshly graveled, like someone's driveway.

And that is when you begin to understand the significance of the walls. Year after year, generation after generation, century after century, the farmers have cleared the fields and piled the stones into rough walls. They run along the sides of most roads and every field, some of them massive constructions ten or fifteen feet wide and as tall as a man. In the winter, when you climb to the top of a hill, you see them running through the woods for miles in all directions under the leafless trees, another sign of land once used and now abandoned. Unless they are a threat to plow blades, the farmers don't remove the rocks any longer, and many of these walls, lacking regular attention, have begun that entropic slide that begins with a missing capstone and ends in a long, low pile of lichened stone.

Standing one late summer day with Raymond among his few rows of vines, American hybrids bred for frost and disease resistance, I asked him what had happened here. He threw out his arm to embrace all his land and much beyond. "All of this land was in vines back then," he said. "Everywhere there is pasture, the garden, the walnut trees, the scrub oak and brush on the southern hillsides, all behind Auricoste and up to the old Domaine de Ladoux, all vines."

And all those terraced hillsides, the pits I had seen in the woods? "That was how they grew the vines on this steep land, with stone terraces, each holding a row or two of grapes. Some you could work with animals, but many it was all by hand." You hoed, manured, tied, pruned, and picked, all by hand. And those pits were for mixing up *la bouillie bordelaise*. That is the patina-green copper sulphate and water mixture which, even today, is sprayed on vines in leaf to keep fungus and mildew away. "Up to 1900," he continued, "we made wine *for export* in Les Arques! Look at the dates on the houses here, especially the finest, up in the village, in the hamlets. You will see 1850, 1860, 1870, 1880. We made good money from our wine. Then came phylloxera."

Phylloxera is a parasitical infestation that slowly destroys the roots of the vine. Over a few decades, it invaded the vineyards hectare by hectare, even as the farmers (for they were farmers before they were vintners) tried every treatment and concoction, every arborist's trick from grafting to preventive destruction in an attempt to arrest the spread of the disease. It first appeared around 1870, and by 1900, the disease had killed more than 95 percent of the vines. That was the first blow, creating a wave of financial disaster and unemployment that led to the first major exodus from the country both abroad and to the newly industrialized cities to the north.

At about the time the vineyards were beginning to produce after being replanted with local vinestock grafted to phylloxera-resistant Californian rootstock, World War I began, a slaughter of mostly the young, mostly the rural, again setting back the recovery of the region. World War II was almost as bad.

Throughout this period, Raymond continued, "the climate changed, it got colder in springtime, and we got frosts. Then in 1956 there came a terrible freeze. It killed all the vines, nearly all of the nut trees. Many of us could not start over . . ." He stops and looks out over the hills, one hand drawing lazy circles in the air as he searches for the words to explain to someone who wasn't there what those times were like. "Our farm we had worked on a small scale," he continued finally, "and it was your whole life. You worked all the corners—corn, oats, tobacco, sunflowers, walnuts, vines *un peu de tout*,"—a little bit of everything. "We had no TV, no cars, no freezer, no refrigerator—and like that no bills to pay, either. You ate only the produce of the farm. Everyone made his own wine, eau-de-vie from the plums and pears. Coffee, oil, and gasoline, that's all you bought. You made your own bread. Each house or hamlet had an outdoor wood-fired bread oven. Later, it was the tobacco, that's what allowed people to stay on the land after '39.

"When my father began to farm, with forty or fifty acres you could support a family. By the time I was a young man, it was one hundred acres, and today it's more! Today there are only [a few full-time farmers] in the commune of Les Arques, each exploiting more than two hundred fifty acres. In 1954, I had planted a half hectare of walnut trees, and I had hoped that that would allow my family to live a little better each year. Then came the frost. We had just been married in 1951, and by then we had the two boys. After the freeze, we had to pull all the trees out. There were five of us farmers, out of ten in the commune, who had to leave the land and find something else to do that year."

Raymond stopped speaking and shook his head at the memory of ripping out those trees, his last hope of staying on the land. He would have been thirty years old then, responsible for feeding not only his wife and two young children, but his parents and the old widow he had agreed to support until her death and whose land he had already begun to work.

"That was also when obligatory social insurance came in for the self-employed," he said after a moment. "That triggered it, really, for those who were already disadvantaged, to have to pay that, too, when you already had no money for the other things in life. And that's when I went to work for the Public Works Department."

Social insurance was such a burden for farmers, not only because it passed into law at a particularly disastrous time for them, but because it was a yearly assessment—income tax, retirement contributions, medical and life insurance contributions all rolled into one—that had to be paid in cash. Even as late as the 1960s, many farmers in marginal agricultural areas like the Lot were not commodity-oriented growers. They grew cash crops like spring asparagus, summer fruits and berries, and tobacco and walnuts, but on a small scale, on a small part of their land. You didn't have a tobacco farm or a hundred acres of walnut trees because you wouldn't have had the resources (fertilizer, seed, and manpower) to make it work on this scale. Cash crops were the exception rather than the rule, and most were usually sold to local processors rather than being turned into something much more valuable by the farmer himself. Cash, as a result, was not something farmers saw much of, and even then only at certain times of year and if the weather had been good, the insects and diseases not too destructive, and sufficient help available both to plant and then to get the crop in. With the two world wars having taken the lives of almost one out of every three men in those generations from this region, this last was more significant than it might seem.

Still, I pointed out, you have a huge garden, an acre or two of nut trees, your vines, a little corn, chickens and rabbits, and you still run your tractor almost every day. He shrugged. "But when we stop, all that will stop, doing for yourself on the side. Our children don't have the habit of it."

"It was all that," Suzanne would add later. "But it was also the times. In the 1960s, the state raised the mandatory school-leaving age from fourteen to sixteen years old, which kept the children off the land for two more years. And then, the thing to do if you were young was to go off to the city, first for more education. And many of them stayed away. Let's face it, there were no jobs here, or only the life of farming, which was not easy."

But, I asked Raymond, what about people like Monsieur and Madame Reix, who had come and started raising goats and making those lovely *cabécou* goat cheeses two miles up the road from us in Lherm? "The Reixes are the exception." Raymond said. "In the Lot, it is true, some of the young ones are coming back. It pleases them to work the land. But it is so hard to take back land that has gone out of use. It was a war for them each year for more than fifteen years. They had loans, had to buy equipment. When too many leave the land, we don't lose quantity—industrial agriculture can always grow more food—but we lose the quality in our food. And *finalement*, the land risks not being worked."

And work they did. From our kitchen window and terrace, we looked out on an open grassy field, then the mayor's garden. On the far side of the garden, his walnut orchard, maybe two acres, ran up the valley, spidery-limbed thirty-year-old hybrids that all grew to perhaps twenty feet in height, their glossy leaves bright against the red of the soil beneath them. Suzanne grumbled about the nuts, which were a lot of work, and the villagers, I gathered, had thought he was crazy to put them in in the first place, since he already had a good job working on the roads, from which he would retire just as the trees began producing. It was just, he told me, that he couldn't bear to see the land go to waste. It was some of his best land, land that had always been worked, and, besides, weren't they beautiful, his trees, as well as furnishing a bit of money on the side?

Beside the walnuts he always put in a bit of corn for the animals, and then, where the land rose more steeply along the far side were rows of staked vines. The late summer and fall were especially busy for them. If he wasn't harrowing under the ripening walnuts, he was spraying his grapes or trimming off the upper leaves so the sun would fall on the bunches below. When we found out he picked his nuts by hand (in my ignorance I had imagined there was some sort of machine), we offered to help. He and Suzanne didn't actually laugh, but they did warn us it was harder work than we might think. We had to wait for the fall rains, at any rate, to swell the thick hulls until they burst, the wind then doing the last bit and knocking them to the ground.

One day in late October, on my way home from picking Lily up at school, we caught a glimpse of Raymond at the end of his drive. We stopped, Lily goggling at the sight of him. He was almost completely purple from head to foot, his shirt and pants stained, his forearms and hands dark with the juice of the grapes he had harvested that day. They were in an oval barrel six feet long and five feet deep and lacking a top, which sat up on blocks in one of the sheds, a cloud of bees and hornets buzzing overhead. He had crushed one load the night before, and this was the last.

"My wine," he told me. "Or at least it will be in a few weeks. Now it's grape juice." He conjured a smudged glass from his workbench, and Lily and I tasted it, smacking our lips approvingly. Is it going to be good this year? I asked. He shrugged. "It is for our table, *un petit vin*." Strong? Although I was no connoisseur, I did know that a measure of potential quality was the degree of sugar in the grapes at the time of harvest, which subsequent fermentation would turn into alcohol. "It's usually eleven, maybe eleven and a half degrees of alcohol," he said. "I'll give you a bottle, but you can't drink it until after Christmas." After a week of fermenting, he would rack and filter it into a smaller, clean barrel that would sit in his cool cellar to be emptied one bottle at a time over the coming year.

Later, both of his sons, who refused to touch his wine, hooted with laughter when I related what their father had told me. "He gets nine or ten percent alcohol if he's lucky," Alain said with a smile. "It's thin, pretty terrible stuff." Gérard just rolled his eyes when he saw his father shyly hand me a corked, unlabeled bottle one day a month or two later.

On a late afternoon stroll, with the first hints of cooler nights to come in the air, we met Raymond and Suzanne by chance in the village center. He was wearing a fancy short-sleeved dress shirt; she a print dress, her hair neatly done. That was what they did on Sunday, put on their strolling clothes and took a promenade, stopping to chat with neighbors out working in their gardens or who were themselves taking the air, people they might not have seen all week. We meandered down in the direction of the restaurant together, nodding to strangers out exploring, perhaps intending to take in the sculptures of the Zadkine Museum before their dinner at La Récré, which wouldn't open for an hour and a half.

Our encounter began as always, Raymond holding out a rough hand. *"Alors, ça va?"* he asked, stepping in close and cocking his head like a curious bird. After the kisses were exchanged and the weather discussed, we looked at the progress the construction crews had made on the presbytery and the five abandoned buildings that the commune, with huge dollops of money from the European Community, is turning into studios, exhibition space, and apartments for a handful of artists to come live and work in Les Arques, more or less year-round. I remarked on the flat, old-style roof tiles they've used, and he told me that it is required now, to use the old style in old buildings in a village like this. These are expensive, flat clay shingles that are more like slates than the long, U-shaped Roman tiles of most new construction. Oddly, however, into the new but quite old-looking roof, ugly black glass skylights had been set. Just across and down the street Raymond pointed out a very old house. "Look at those upper windows there, you can see what they used to do."

The house has two small dormers whose front vertical face is formed of several dressed stones, their great weight supported by the foundation wall on which they sit. (This is why these old-style dormers are always set so low in the roof, as well.) The uppermost block has a triangular top, mimicking the little piece of projecting roofline it supports. This is called a lucarne, and they often have designs incised into the stone on the bottom corners and the upper extended triangle. Here, the design is a semicircular sun spreading its rays up top, with a simple geometric, almost Celtic-looking motif in the two lower corners. "That is typical for the region," Raymond told us. Later, I would read that these sun signs are thought to be a holdover from pagan times.

"Here," Suzanne said as we passed a house with an arched passage into an inner court set into one side of its lower story, "is where the smithy was." A long time ago? I asked, thinking they meant eighteenth century. They both laughed. "I don't think he retired until almost 1960," Raymond said. So he made all the beautiful hand-forged hinges and shutter catches and door handles I had seen on so many of the houses. "Oh, yes. But mostly he made shoes for the horses and the oxen and shoed them, right there, in the passage," Raymond said. "Most of us plowed with oxen right up to then." And the first tractor in the area? "*Alors*, that would have been that farmer over in Lherm, what was his name?" Suzanne couldn't remember, either. "But it was probably 1955, around there.

"But it wasn't until 1965, 1965!, that running water arrived in Les Arques," she continued. "*Mon Dieu*," she exclaimed, frowning, "but I remember hauling those buckets, morning, noon, and night, fifteen gallons a day, to the spring down below us and back up the hill. And we had electricity much before that, but it was for lights, mostly, because there were no appliances, certainly no washing machines. Dirty clothes we hauled sometimes twice a week to the *lavoir*, the washing place, those flat stones set into the bank of the stream by Auricoste. Can you imagine? With two little ones and diapers to clean. There is a lot about that time that was not so charming or pretty, just hard, hard work."

Outside the village, we started down the track through the fields to Auricoste, and the mayor pointed out the path he'd just cleared. What appeared to be just another grown-over tractor path was actually a *voie communale*, or a road belonging to the commune. No one paid Raymond to mow it or keep the weeds and nettles back from its verges; he did it because it looked better, and because he knew people liked to walk there.

"Have you ever gone down this other track, around and down toward the château?" Suzanne asked. "Are you busy? No? It's worth it!" The path wound down, past garden plots with beans and tomatoes and cabbages, down to a rivulet running beside a small barn, across from which was a half-acre field of turnips grown for animal feed.

Raymond told us that the empty house at the end of the Auricoste road, formerly Zadkine's guest house, had been sold, to a businessman as a summer home, apparently, and talk moved on to local real estate, who was buying or selling or which house had been vacant for decades because the heirs couldn't agree to restore or sell up.

There was a wrinkle in French law, I learned, such that, unless special provisions were made before death, a father was obliged to pass his estate in equal shares to his wife and children. (The wife, even today, does not necessarily have the right even to stay in the family home for her lifetime unless she owns it already, which leads to regular newspaper stories of greedy children selling the house out from under elderly Mom to collect their cash.) On the other hand, there is also no disincentive to holding property, especially if it is used only as a second home. Property taxes, on the order of two hundred dollars a year on a hundred-thousand-dollar property, are so derisory that there is no real carrying cost.

Finally, maintenance is also not a significant issue. The electricity and plumbing systems of these old stone houses are often of the most basic sort, as are the amenities. Many remain as Grandma and Grandpa left them, heated by huge wooden fireplaces and with simple kitchens consisting of a gas cooker and a stone sink, with the privy out back. They do need the occasional roof and to have their stones repointed, the former perhaps every forty years and the latter once a lifetime. Aside from that, they soldier on admirably with a slap here and a dash there, unlike so much modern construction.

We climbed a gentle dirt road, at the top of which was a stand of old-growth chestnut trees in their final years of life, some six or seven feet in diameter. Many had lost their crowns, most were rotting from the inside or had been taken by the hurricane-force winds of Christmas 1999, which were responsible for so much of the deadfall we saw littering the woods. Around the corner was a château in the country style, more of a huge farm spread with two real stories and towered *pigeonniers* at its corners and vast barns and outbuildings. "Some foreigners bought it," Raymond said, "people from Paris. They come and buy all the biggest places, high up and with a view. But these big estates, especially with a lot of agricultural land, the state has the chance to buy them first. Sometimes they do." He shrugs. "Sometimes they then break them up into smaller parcels and resell them just to get their money back."

The vast property, surrounded by ten-foot-tall electrified fences and amply marked with warning signs, is quite forbidding. In this corner of the country, where public footpaths take you past most houses and where right of access is defined not by law but custom, the little château is the exception and an egregious one at that. You cannot approach the house, or cross the property, and inside the fence a herd of fallow deer graze, an extravagance in this poor landscape somewhat akin to the hunting parks surrounding the private great châteaux of yore. Of course, being fenced, the whole spread is also off-limits for those other great country pastimes—gathering mushrooms and hunting on other people's property.

"Do you know them? Do you see them in the village?" I asked. Raymond looked at me and laughed! "We know who they are, of course, but we don't know them. They don't vote, I think. And they would never show their faces at the village fête! Oh, they'll come to the monthly Mass at the church, yes, but the fête or other village functions? No. They don't make the effort to integrate themselves like some who come here." As we passed, he cast a glance back at the fenced fields now given over to grass where once crops grew, a sour look on his face at so much wasted land, and for what? So some gentleman who is not even a farmer can look at his exotic deer four weeks a year.

He related the story of another incomer, a woman from up north, a retired university professor who bought the farmhouse on the hillside opposite the château many decades ago. "She lived very simply, planted a garden. She used to tell me, 'The garden nourishes me, and, in the fall, the woods feed me!' She was talking about cèpes and girolles [porcini and chanterelles], the mushrooms which she gathered. She died very suddenly at age ninety-four, and we all went to her funeral."

I remarked that maybe the children of these new landowners would take to the lifestyle of the country a little more. *"Non, non,"* he said rather wearily, a voice of long experience. "I doubt it. More likely there won't be enough money to keep the house and the land up and so it will be sold again."

He mentioned the Domaine de Ladoux, the large property with a small château whose slate roof you see on the way in on D150 from Maussac and from the village cemetery. "A monsieur bought it. From Paris, maybe. He's a writer, too, I think. But mostly he's an environmentalist! He's let everything go, the fields grown over, trees and scrub taking over." He shook his head at all that lovely land lost to some idea he understands but doesn't much sympathize with in this circumstance. "Even the château, I don't see him doing the things that need to be done."

By this time the light was just beginning to fade, and we were coming up on La Mouline, what we had already come to think of as our place. I thanked them for the enjoyable hours, apologized for all the questions. They invited us to dinner a few weeks off.

• • •

Suzanne's kitchen, except for the thick stone walls, the *cantou* fire-place taken up by a Franklin stove flanked by two rush-seated chairs, and the shotgun over the doorway, could be out of a 1960s American sitcom, all avocado and rust appliances and bright table-cloths and lace curtains at the windows, large TV set in front of two comfortable chair at the far end. When we sat down at the table in the dining room, however, we were squarely and solidly in nineteenth- or even eighteenth-century rural France. The walls, of rough-plastered stone, were lined with a rustic walnut tallboy and chestnut linen cupboard almost Shaker in their simplicity, the ceil-ing crossed by stout beams stained dark over a fireplace into whose stones the date 1767 had been carved. The table was set with a finely embroidered lace cloth faded to a rich cream, the etched wineglasses squat and tulip-shaped, the silver heavy and simple and bereft of ornament.

Suzanne set tiny, tangerine-fleshed Quercy melons before us, their centers filled with a few tablespoons of sweet muscat. One of Suzanne's whole duck foie gras, thick slices pink to tan to brown, served cool with bread followed. I had always thought the consis-tency of foie gras resulted from some sort of processing, but what you see is what you get, in fact. It had been deveined, dusted with salt and pepper and canned in a hot-water bath, but otherwise it was in its natural state, a firm, completely smooth and uniform roughly cylindrical log. Although there is in its taste the distant memory of liver, the fattening process renders it something alto-gether different, much more subtle, much lighter, almost like a fla-vored butter on the crusty bread.

We sat, ate, drank, murmuring sighs of approval, Suzanne smiling from the far end of the table, insisting we take another slice, of which I managed only a half, for, also like butter, foie gras punishes immoderation. I thought of the foie gras we had offered Raymond and Suzanne. It was a paltry offering, nothing at all like this.

"That was different!" Suzanne said. "That's more a commer-cial production, and that was a little foie gras. The bigger they are, the better they are, the more solid the flesh. And of course, much better in the mouth."

The melon and foie gras were followed by *une croustade de canard*, a big pastry pie stuffed with chunks of carrot, turnip, potato, and duck confit. Each slice was served with a small thigh and drumstick of the preserved duck, its very dark, almost purple flesh peeking through the mound of vegetables and crust and dark gravy on the plate. The dish is seasoned with salt and pepper and the merest hint of garlic, but it is pure and simple the unadorned vegetables, the rich crust, and the slightly gamy taste of the duck that come through in each bite. I had seconds.

"It's a peasant dish." Suzanne explained almost dismissively. "This is the kind of thing we ate when we were growing up; then, in the sixties and seventies, we didn't eat it anymore. But now, we eat it again sometimes, *un peu de nostalgie*. Like *la mique*. You remember, Michelle, when Alain first brought you by?" she asked her daughter-in-law, Alain's wife, both of whom were eating with us that night.

"Yes, you made it the first time I came."

"And you didn't like it, I remember . . ." Suzanne said, as only a mother-in-law could.

"*Si, si!* It was good!" Michelle protested.

La mique is similar, a slow-cooking bouillon of vegetables and meat, onto which is poured a batter of egg, baker's yeast, and flour cut with goose fat and a little milk. It rises as it cooks, gets flipped once, and is served so that each plate gets a ladle of the vegetable and meat in their cooking juices and a slab of the rich, cakey dough. This is one of the dishes once used to measure the proficiency of a housewife in the kitchen, and heaven forfend if your *mique* sank into the stew rather than rising on top.

A charlotte with peaches in a custardy cream followed, with Champagne to toast our health. We had been two and a half hours at table, and my wife and I would both roll into bed that night vowing to eat a little less next time and walk a lot more between meals.

"When we first heard you were coming," Suzanne said over the coffee, "we didn't know what to think! We thought, maybe they're very different from us, we see how Americans live on television, they *must* be very different. And when you saw us that first night, when we were bringing in the potatoes, Raymond driving his old tractor, I thought, they're going to think we're a couple of old peasants!"

The odd thing is, they are a couple of old peasants, neither with more than an eighth-grade education. They've gone to Paris a few times in their lives, and they've been on an airplane once, to Morocco last year. And yet they are peasants who can tell you that Bernard Pivot is the host of *Bouillon de Culture*, a weekly hour-and-a-half-long TV show in which various authors, politicians, artists, and other people of culture come to discuss, with all the considerable erudition the French intelligentsia can muster, their work, world issues. It is a program consequently about five times more intellectual and demanding than the best of even PBS. And it airs on Friday nights during prime time!

Paysan literally means a person of the land and generally has much more positive connotations than the word "peasant" in English. Here, it evokes the farm, the earth, and someone who labors with his hands to produce for your table, someone who knows the old ways, who follows a life dictated by weather and the seasons. While city dwellers may deliver it with a wagging finger, hinting at the vast cunning and hidden agendas of their country cousins, out here it is who you are, no different from saying you're a doctor or a shopkeeper or an electrician.

Suzanne, *paysanne* though she may be, was the one who explained to me the more subtle aspects of the August 2001 standoff between Lionel Jospin's Socialist government and striking truck drivers who had blockaded all the oil and gas depots, paralyzing the country. Jospin, Suzanne continued, had just refused to negotiate further, to make further concessions, a move he'd taken only because the truckers' union was a weak and disorganized one, unlike the railroad or civil service employees. In essence, Suzanne told me, he was counting on the fact that, after a few days, public opinion, which had at first unanimously supported the action, would turn against it, forcing the truckers to back down. She was correct.

Raymond is a very quiet guy, and, with his slow smile and rough speech, it is easy to take his measure as a simple farmer and no more. Yet he's also someone who has seen the changes large and small that have overtaken his small corner of France and thought about them. "Is it a good thing that foreigners are coming in and buying up abandoned properties?" I asked once. He stopped to think, as always. "Yes . . ." he said, almost reluctantly. "It's better when the houses aren't falling down into the street."

But later, when we were talking of the neglected Domaine de Ladoux, he showed a bit more of his hand. "He's let the land go! That used to be the richest farmland in Les Arques. When I was growing up, and you got out of school, you went up there if you needed work. In the vineyards, or the fields. They used to grow wheat there, down below, those large fields behind Auricoste. Wheat! And rye. It was worked until the 1960s, when the farmer who owned it died," he said with a touch of anger. "Now, the environmentalist! Trees growing up in the fields, walls falling down. He doesn't even seem to be maintaining his house as he should," he finished bitterly. It was the only time I'd ever seen Raymond even the slightest bit angry.

For us, the Lavals were a lifeline, our first friends, gentle wise elders ready to share their lives, their knowledge, their world with us, to open our eyes and, especially important for our daughter, their hearts as well. One evening later in the fall, Lily was playing out on the terrace when the mayor drove by on his tractor. I was making dinner, and saw from the kitchen window that she had stopped and was watching him intently as he drove slowly from one end of the garden to the other. "Can I go ask Monsieur le Maire for a ride?" she said, poking her head in the door. "Go ahead," I said, "just don't get in his way." I was surprised, for, though she had been in school for two months, she had always insisted that one of us do the talking for her.

She crossed the meadow, the grass almost as tall as she was, then paused at the edge of the garden. Raymond stopped, a hand cupped to his ear, then helped her onto his lap. It was my turn to watch, amazed and not a little bit proud, as they paraded up and down the field, Lily chatting nine to the dozen and the mayor nodding and answering. Even after a few short months of acquaintance, I knew that in this older couple and their extended family, we had stumbled upon a little world of good. Life here seemed almost too slow and simple at first. As we settled into the rhythm of the country and its inhabitants, however, it was turning out to be amazingly rich, complex, and full of surprises like Monsieur le Maire and his family.

FOUR

A DEAD SCULPTOR

This village is dying. But it will not die as quickly as the others.
—OSSIP ZADKINE, SCULPTOR

One wet morning after we had been established in Les Arques for several months, the dog and I made our way up to the village by the path from Auricoste. Isabelle, her black coat shiny with the cold rain of fall, ran on ahead, dashing off into the woods whenever her nose hit an interesting scent. She knew the village well by then, had romped its lanes many times, been patted and cooed over by the old ladies who post themselves each day like sentries in their kitchen doorways waiting to check the credentials of the passing world. I came out onto the narrow street behind the church to find Madame Carrié on duty, tending her potted plants. She was in uniform, a dark formless sweater over housedress and apron, mules on her feet, her tiny yet rotund form bent over the thirty or forty potted plants that seemed to consume most of her time in summer, fall, and spring.

"*Bonjour, Madame!*" I called out, having earlier discovered that her hearing was not what it had been. "Another gray day, no?" Her back was turned, and she wheeled abruptly, peering out at me.

"But who are you?" she asked suspicously. "I don't know you, monsieur!"

I was a little taken aback, as I'd been passing *le petit bonjour*, chitchatting, with her on our frequent jaunts through the village. A week earlier, I had stopped by to admire the enormous parsley plant, luscious green, aromatic, and the size of a small shrub, that she had grown in a clay pot and that had been for weeks the talk of the village. (That the parsley—and the repainting of the Virgin, who appeared to have shopped the fall sales and come home with new robes of brilliant robin's-egg blue and eye-dazzling white— was the big news here gives some idea of what passes for local excitement.)

I pulled off my hat and came closer, and her eyes brightened. "Ah, yes, you. I didn't recognize you without your dog." Isabelle had come back from inspecting the neighbor's bushes and was whacking her tail against my leg, looking from me to Madame Carrié. Always a polite dog, she was waiting for an invitation to go say hello, which she duly got, bringing a smile to the lined face and a shower of the baby talk that the French reserve for kids and dogs.

I sometimes had the impression that the villagers were more comfortable talking to my child and my dog; it took me a few months to figure out that they were having a hard time situating me in their mental landscape. I was a stranger, but one who had stayed on after the others had left. I spoke good French but didn't seem to work at any job they recognized. I popped up all over the village and asked a lot of questions—and noted down their answers. In another era I might have been taken for a spy, or worse, maybe even a tax inspector, and run out of town.

Happily, the grapevine had long since passed on the news that an American was coming to write about the restaurant, but they had more difficulty understanding why I was so interested in the bigger picture—their lives, their village, their stories. "I thought you were only here for the restaurant. No? You're interested in the death of the vines (or the church, or Zadkine, or indeed any concrete topic I brought up)?" they would ask. "You need to read that book, Madame Auricoste's book. I don't know anything. I'm not an expert. I hardly even went to school!"

While "that book," Françoise Auricoste's *Les Arques in the Quercy: Valley of Iron, Valley of Art*, was very useful as a basic local history, any text would appear rather dry and colorless in comparison to what I saw before my eyes every day, to what I learned in my meanderings and conversations with the villagers. Madame Auricoste, though undoubtedly a good person, lives in a distant city. She probably does not make her own wine and foie gras each fall, cannot be found down a distant lane cutting wild dandelion greens to fatten her rabbits, or out in a smoky back shed distilling a few liters of plum eau-de-vie on the side. She does not spend a frigid December morning stuffing the intestines of a freshly slaughtered pig with its coagulating blood to make *boudin*, the black blood sausage, or a July afternoon bent over the vines, pruning exuberant green growth to let the sun hit the grapes.

All of these things go on in and around Les Arques each in their season, with some villagers still living in many ways as their parents and grandparents did before them. It is this knowledge in doing, the savoir faire seen in the gnarled hands and mudstuck boots, the burgeoning back gardens and pens of fattening ducks, that strikes such a resonant note to someone from the Internet-crazed, technology-driven, "modern" world of middle-class America.

One of the advantages of being an outsider, a stranger, is the perspective this status gives you, almost as if, in setting you apart by treating you differently, the locals give you permission to consider them more closely than if you were one of them. There are no expectations of you, and you are allowed more liberties, at least until you prove yourself unworthy. The Arquins, as the inhabitants are called, are certainly, like the French on the whole, very private people, wary of outsiders perhaps from a long (and continuing) history of being invaded by them.

From a purely peasant point of view, for instance, unless your house is for sale, or the family is celebrating a marriage, welcoming a newborn to this world, or ushering off a departed elder to the next, there is no reason for others to see the inside of it, certainly not beyond the kitchen curtain. Inviting others over for a casual meal (and forget spontaneity altogether) just doesn't seem to happen much outside of the traditional gatherings of Christmas or Easter, when those you do invite you generally went to school with—elementary school. You might see the same person on the street every day, even stop to chat of this or that, but an invitation for more than a cup of coffee just isn't going to be forthcoming. Needless to say (and to our great frustration), the concept of the child's play date—informal, spontaneous, and wildly American—hasn't yet quite caught on in this part of the world.

Outside of Jacques and Noëlle and the Lavals, all of whom are atypical in this respect, we were invited into private homes only rarely in the course of a year. The same hesitation held true in how they felt about coming around to our house. With Suzanne Laval, it was like coaxing a wild bird to take food from your hand. She would stop by usually to give us eggs from her hens or produce from the garden or to check that a workman had passed, would hover at the top of the steps leading to our front door. "Anyone home?" she would call out, even though she could see us through the glass doors.

At first, she would perch only momentarily in the kitchen doorway, clearly wary of intruding, refusing all offers of food or drink, and visibly anxious to be on her way. As this was usually in the late afternoon, my wife or I was generally in the middle of making dinner. She, and her daughter-in-law especially, was clearly dying to know what and how we ate (as were, in Cazals where we shopped, Lydia the greengrocer, the butcher, and Madame Miguel who owned the little supermarket), but both in the beginning equally reluctant to come right out and ask. This was not from shyness or reticence, but respect in its most elemental and ancient form—respect for your castle and the privacy attached to what goes on within its walls.

On the other hand, the street outside your house, your front yard, even your doorstep, these were all neutral zones where your neighbors felt free to penetrate, there to pass on the news of the day. Almost every afternoon when I drove down to the bottom of the lane past the mayor's house on my way to pick up Lily, he or his wife would be standing at the end of their drive beside one neighbor's car or another, or on the doorstep of Christiane, who lives across the street. Driving the local roads, even the busiest, we often came upon two cars traveling in opposite directions, their drivers stopped, jawing away through open windows. One would reverse to let us pass—then pull right up again and continue the conversation.

Soon we were halting, too, having realized that rocketing by (we took to rocketing pretty quickly, as well) without at least a word or two was not only rude, but a missed opportunity to hear the news and participate in the local life. That Christiane had a terrible toothache and needed a lift to the dentist because her son was out of town; that Yvonne, also housebound, would appreciate a round loaf and a small chicken from the Sunday market. We would never have known to offer without pausing for a lengthy hello, would never have then subsequently reaped the small kindnesses and greater intimacy that followed on these little favors, which were not, by any means, all one-sided. Favors are the small currency of living in the country, and it always helped to have a few in your pocket for a rainy day.

Though we soon came to know quite a bit about country etiquette just from observing and imitating (with generous dollops of advice from the Ratiers and the Lavals), there were also those kindly souls so graceful that we never knew until afterward that they had helped us avoid deeply embarrassing gaffes. Toward the end of October, I had noticed that the markets, the groceries, and gift shops little and big, all were bursting with displays of the lush-

est, fullest chrysanthemums, dizzingly colorful and tempting. We
had been invited to Noëlle's traditional family gathering outside
Toulouse, and I had been deputed to buy the bread-and-butter gift
for her mother. At the greengrocer, a mammoth pot of bronze
mums had caught my eye, but Lydia Bouysset, the proprietress,
kept steering me away from them. She probed delicately, "And
Madame is perhaps an older woman? A relation? She lives in the
Gers [a neighboring *département*], you mentioned?" and so on,
each inquiry carefully and respectfully phrased, before finally
guiding me toward a bright bouquet of fresh flowers. Little did I
realize that Lydia had only been trying to help me avoid offering
flowers appropriate only for a dead person, and to Noëlle's
mother at that. Potted mums, acres of them in every hue, are what
you see in the cemeteries at Toussaint, the All Saints' Day holiday,
when many families gather to pay their respects to their ancestors
by cleaning and adorning the family tomb.

Quite early on in our year, I had stopped by a local *gîte*, a
guest house, one noontime to reserve rooms for visitors who
wouldn't arrive for months. The summer sun pounded down on
their terrace, the middle of a season of blinding light and soporific
heat to which we hadn't yet adapted. As I fretted over dates and
leafed through the moist pages of my weekly planner and tried to
work out who would take what room for how long, the owner and
his wife exchanged a quiet smile at my expense.

Guy, the genial owner who had retired here from Belgium with
his wife a few years earlier, held up a beefy hand. "Monsieur, don't
worry! We'll figure it all out. You must take up the rhythm of the
country." It was good advice, if hard to follow at first, the rhythm
here being very slow indeed. (Also in the first weeks, we had taken
a broken television into Monsieur Blanco's repair shop, and he had
kindly lent us a replacement. At first, every few weeks when I ran
across him in Cazals, I would ask if ours had been fixed. *"Eh
biehng, non!"* Nope! "Is there a problem with the loaner?" he
would ask with concern. "No? Well, then . . ." We never did see the
television again.)

From the villagers' point of view, ours were just the latest in a long line of strange faces come to call, and some to stay. I often wondered if they perceived the irony in their present situation, that the fate of their home, their place, their beloved streets and church and houses of stone, should have come to rest on the backs of these same strangers, three in particular.

"If this village has come back to life," says Patrick Cantagrel, the new, young mayor who replaced Raymond Laval in 2001, "it is due to someone who himself came here as a foreigner and chose to work and live here. I'm thinking of Zadkine. And he certainly left a very interesting inheritance to the village which indeed has permitted it to remake itself twenty years after his death. The Zadkine Museum has been the driving force of much that has happened here since its opening in 1988. And if there is a Zadkine Museum, it is of course because we had Zadkine."

It is hard to overstate the importance of the accidental coming of Ossip Zadkine to Les Arques, and of his presence today, in the official and unofficial life of the village, which is all the more remarkable since he is an artist largely unknown outside Europe. I certainly had had no idea who he was when I arrived. In the period after the war until his death in 1967, he was widely identified as France's greatest living sculptor. He won the Grand Prize for Sculpture at the 1950 Venice Biennial, exhibited widely around the world, had his works purchased for state museums in France, Belgium, Holland, and the United States to name but a handful. In Paris, his former studio on the rue D'Assas houses a museum exhibiting many of his works, all of which he left as a legacy to the French state on his death.

In Les Arques there is his museum, of course, and his sculptures, some of them massive, the wooden *Pièta* and *Crucifixion* (which everyone calls "the Big Jesus") in the church as well as a group of bronzes in the courtyard separating it from the museum. (Photographs of his art can soon be seen on the Zadkine Museum website at www.zadkine.org.) The easiest thing to discern is the Cubist influence of his early years in Paris between the wars. In the two rooms of the museum, one the size of a large living room and the other a spacious, cathedral-ceilinged expanse, there are examples of his work in many media. What catches the eye, however, are the very large figures—Diana and her dog, Orpheus, Daphne—almost crudely conjured of misshapen parts, mixed perspective, and odd volumes, a very raw, almost primitive, hyperstylized appearance neither wholly abstract nor wholly representational.

In the guest book, visitors write: "The stuff of nightmares—wonderful, alarming!" "Why so many Orpheuses?" "Dislocating the body is not the answer." "Zadkine seduces, moves, welcomes, enriches, offends us, is perverted, subverted, inverted, and diverted." "*Merci bien*, Z.! You have given us such beautifuls!" "Ridiculous! You really don't put up museums for just anything. But okay, he still married well."

The first words that came to my mind after spending some time in the company of his pieces were "Bad Picasso." I understand the intent, I appreciate the labor, but his work leaves me cold. Cold, too, in that he appears, in his sculpture anyway, to be bereft of the least shred of humor and warmth, and very, very distant from human emotion, which is a bit odd since he spent so much time sculpting people, mythical though they may have been. What struck me most about his religious pieces, in particular, was that their attraction began and ended with their intentional brutality, rendering his Crucifixion, to my perhaps Philistine eyes, nothing more than some guy on a cross.

There are his house and studio, now privately owned, mostly uninhabited, and hidden behind a high stone wall and imposing gate (but still to be seen for those who don't mind peeking). There is his guest house, just down the road out of town, once derelict and now being restored after a recent sale. The sculptor's art is celebrated in Zadkine's name every summer with a visiting exhibition of the works of another, usually contemporary, sculptor, an exhibition whose opening draws every muckety-muck for miles around. And there is Les Ateliers des Arques, the organization founded by a handful of young Arquins who tapped the capital of his name to bring a small group of visiting artists, mostly sculptors, to live and work in the village first in 1988, a program that continues to this day.

Zadkine is still very much alive in the memories of the locals, too, many of whom are old enough to have had long and sometimes meaningful contact with him. Everyone has a story or two, a kindness done, a hand extended, a moment when the gulf separating peasant from urban gentleman artist was bridged, even if only by a smile or a *bonjour*. "He traded me a white shirt once," Raymond told me one day. "It had a high collar, high like this . . ." He spread thumb and forefinger several inches, then mimed long sleeves and doing up French cuffs. He looked me in the eye, his face creased in puzzlement at exactly to what event one might wear such a thing. "It was during the war, and people were hungry. No one had any money. I gave him a young pig. I can still see him walking away, with that pig slung over his shoulder. I never did wear that shirt."

"He was very nice, very kind." Elise Ségol recalled as we sat in the tiny, immaculate kitchen of her house across from the restaurant one day. "He would as soon speak to a cat as a dog in the street!" (Meaning he could and would talk with anyone of any social class.) She leaned back in her chair, a sturdy woman with square features and dark eyes, her gray-white hair cut short, looked down at her work-roughened hands with their short, swollen fingers on the table before her.

"For twenty-five years I worked for him, or for his wife. Cleaned house. People left him alone, mostly. He spoke well, and he was quickly adopted by the village. His wife . . . less so. He wore a hat, always, and an overcoat like Columbo." She chuckled at the thought. "And he smoked a pipe. He gave us work, he did, paid us to work the orchard, the garden. They employed a lot of people here and there. He'd hire the men to bring up those huge trunks of trees out of the woods. His wife was a painter. Not at all like him, less well . . . appreciated certainly." She sniffed, raising her nose a single centimeter in the air, giving me to understand that Valentine Prax, the wife in question, was a bit of a snob. "She was not the kind of person he was, of the land. But we didn't know him, not really. He had us come to see his work sometimes. He explained things." She shrugged.

"Fifteen years ago? There wasn't anything left here! The restaurant didn't yet exist. The school was down to twelve students, and I cooked for them. Then the school closed. Zadkine, the museum, it brought people here, visitors, to see the church, the village. It's all due to Zadkine, for they wouldn't have come without him."

Intimate is not a word that seems to fit any relationship he had among the Arquins. In nearly every photo of him at the museum and in the books about him, he is alone, usually outdoors, a short, rather awkward-looking figure with a shock of white hair, jutting nose and brow, pipe clenched between thin lips. His face has a sunken, ravaged look that, together with a thin neck jutting from large collars and oversized jackets around too thin shoulders, makes one think he did not have much to eat for long periods of his life. And yet he seems, at least from the long lens of today, to have been quite well regarded by the majority of villagers, barring a few unpleasant years during the war.

Blanche Delfort, eighty-seven years old, and her daughter Simone, as near neighbors, had more contact with him than most. Blanche moved to Les Arques with her husband from an outlying hamlet in the late thirties, shortly after Zadkine's arrival at the same time the new school was built. I asked her what Zadkine might have seen when he arrived, what the village looked like, how people lived.

"Then, there were a lot of houses that people lived in which don't exist any longer. I don't know the exact number." (The population was probably around three hundred then.) "There was the blacksmith, Marroux, who had the smithy where the arch is at the bottom of the main street. He did a good business, shoeing horses and oxen right in the street. His wife was a dressmaker..." As well as the many who fed themselves from the land, there were two masons, a ragman, day laborers, a goatherd and cheesemaker, a seamstress, a roofer, and a tilemaker.

Blanche's husband was a mason—"he helped build the school"—while she had a small grocery selling mostly dry goods in her kitchen. "It was not very important—coffee, sugar, salt, oil, flour, pasta. At that time we were satisfied with how we lived. There was no misery; we had chickens, ducks, pigs, and we lived from that. We didn't have a freezer, of course, but we pickled and preserved, and we kept the meat packed in fat in earthenware pots. There was no butcher shop here, no need. For *les gros travaux*, the grain harvest, the grape harvest, everyone helped each other. There was a lot of communal labor. But we didn't depend just on our bit of land because my husband had a job, had money coming in from his profession as a mason. We were not among those who had only land and only a living from that land.

"There was no bakery. We all made our own bread. The baker came through on his horsecart, but most did it themselves. My husband built a bread oven out back, but there was another, a communal oven right behind the Mesmains' house. Each would bring his own wood, and the dough. For us, I made our own of good flour, for two weeks at a time, huge round *miches*, loaves big and round like that." She spread her arms as wide as she could to still make a circle.

"Zadkine was easy to talk to, with, bless him, a kind word for all. He hired a village woman to clean his house. He was very kind like that. We worked his land for him. When he came alone, he would pass by and pick up his soup to heat up at his house when he wanted it. He had chickens and cats, and we took care of them when he had to leave during the war and whenever he wasn't there. No, he would never invite us to eat. You didn't do that! People were less open back then. He lived very quietly, liked to walk, a quiet life."

"We had the keys to his house," Blanche's daughter, Simone, added. "We saw a lot of his work as it was being made, could see how his pieces were coming because that's where his cats lived, in the atelier. We lived very close, right next door to him almost, and we were not aloof from him like so many others in the village. The people here did not know him, and did not understand his work. They saw it and they respected it and him as someone who worked hard—"

"But we didn't know anything about it," Blanche broke in. "What he would say to us always was, '*Il faut le sentir, il faut le sentir'!*" You have to feel it. "How many times I heard him say that!"

Zadkine, like so many before and after, quite simply fell in love with Les Arques at first sight on a day in early spring, 1934, and especially with a property for sale perched high up on the eastern edge of the village. In *The Mallet and the Chisel*, he writes:

> *Early in the spring, Valentine and I made our way to the village [and] found a large house—and largely dilapidated as well—with an enormous barn in front, a sculptor's dream! The house made one think of an old woman, who, during a long life, has put up with plenty of unhappiness and has many distressing stories to tell. The stone staircase of the tower led to the first floor of the building, and what an unforgettable tableau this first floor offered! A tall and wide armoire reigned imposingly in the middle of a vast room lit by two glassless windows; a single bed sat, long barren of any intimacy . . . [T]he incessant fluttering of two swallows busily building a nest brightened the space, and an optimistic sun streamed in. It was magnificent!*

What attracted him was not only the space for living and working, but more importantly the country setting so rich in stone and wood and so far from the noise and dirt of the city. Throughout his writings, and in the recorded oral history one can listen to at the museum, he speaks of the deep and very personal relationship he felt with the Lot, with the trees and stones of Les Arques. He wrote of those first impressions:

*Looking out the first floor windows, I could see the porch of the
barn, bright and open, which smiled at me. Further, surrounded
by a stone wall, a tall cypress tree . . . whispered to itself with its
thousands of diminutive, lacy leaves. Even further, out in the
countryside, thick forests climbed hills on which the houses and
farms stood out like huge, shining stones. Still further, forests of
pine and chestnut spread out under a vast sky of distant and inde-
finable blue.*

When a person becomes an institution, as Zadkine has become
in Les Arques, the harder edges of his life tend to get softened, the
burrs smoothed over, the difficult times seen in the rosier light of
retrospection. If the village has unconsciously buried the less pleasant aspects of the Zadkine story along with its electric and telephone wires in order to present a more perfect image to the world,
it is understandable. For those "unpleasantnesses" came mostly
during a time of great privation and upheaval, the Second World
War, a time still very much alive in the memories of the two oldest
generations, one then coming into early adulthood and the other
still children.

Zadkine had been living in Les Arques for less than ten years when
the bitter winds of anti-Semitism aroused by the Second World War
blew through the village. "The atmosphere was strained, very difficult." Simone explained. "It was a general malaise, a time—"

"A very bad time, bad, bad." Blanche broke in. "I had two
brothers called up, one mortally wounded, the other taken prisoner. My father-in-law taken prisoner, too. All the young men
went to the front, or were taken prisoner, or went off to the resistance in the south, so there were few men to do the hard labor. No,
I can never forget those times."

"—you no longer could trust certain people," Simone continued. "[It was] a time where there were conflicts of opinion, and
you didn't want to confide your own thoughts to others for fear of
how they would be taken. It was a time of repression, remember,
and so relations were delicate between people."

First, Zadkine's name appeared on a list of Jewish artists published by a "patriotic" Parisian journal. Next he was denounced by a local bourgeois, the owner of Ladoux at the time. He fled to America, leaving his wife, Valentine Prax, to watch over their home in Les Arques. Valentine writes of these times in her memoirs, *With Zadkine*:

> I will not speak of the wickedness and cowardice of certain persons of this village of Les Arques, which we had adopted but which, to judge from their behavior in these circumstances, had not adopted us. I will not therefore speak of the misery which I endured at the hands of the civilian militia and of those who denounced us. To think that some would do anything for the color of a single banknote!

The period of the war is something so sensitive that people who lived through it still do not like to talk about it at all, those moments noble or less so. The entire southwest was the cradle of the *maquis*, who harried the Germans so effectively—and suffered so much for their efforts—in the later stages of the war. This area is also where the Allies dropped the dynamite to these resistance fighters to blow the rail bridges linking Toulouse and the north, delaying a Panzer division on its way to reinforce Normandy after D-Day. And one day walking near the border of the commune near Lherm, Raymond pointed to a stone house. "There," he told me, "during the war a clandestine radio operator lived. He appeared, and we were told he was someone's relation. One day near the end of the war he vanished."

Though I put questions about these times as cautiously and gently as I could, I was still surprised at the deep reservoir of bad feeling they seemed to tap. The villagers' faces would close up, their eyes look away, and the sentences would emerge short and shorn of detail, bare outlines of events still quite alive in their memories.

When I asked the natural question, if Zadkine had been so well liked, why had he been denounced?, the response was as unsettling as it was curt. "That person," Simone told me, not even willing to utter *that person*'s name but instead jabbing her finger emphatically out the kitchen window across the valley toward Ladoux. "That person had his own opinions about Zadkine." Blanche nodded, muttering, "*Oui. Oui, oui. oui, oui, oui! C'était horrible, affreux, la guerre!*" as if to emphasize her daughter's meager words. (That person was Monsieur Verny, the largest local land owner and apparently someone who fancied himself of higher station than the villagers.) Zadkine pulls no punches in his recollections, *The Mallet and the Chisel*:

> *A wealthy resident of Les Arques had set about stealing our dear old house . . . This man wanted to help himself to the property he had left in France, this M. Zadkine, this half-Jew gone to America, and to the property of the wife, this Madame Zadkine, who had stayed [behind]. [While a friend in Spain searched out the documents necessary to prove that Valentine was not Jewish], Valentine was suffering in Les Arques, a frightened woman surrounded by atrociously indifferent neighbors, a woman forced to endure humiliations, visits from the gendarmes, and harangues from that miserable Monsieur V., the self-proclaimed true Frenchman, who tyrannized her by calling, ever more loudly, for her dispossession of the house and all its contents.*

A settling of accounts seems to have taken place after the war, in this instance not nearly so brutal as some, the SS collaborator summarily executed in a hamlet outside the village, for example. "After the war the village was divided over what to think of that person." Simone said. "Many people avoided him. That same animosity he had had to Zadkine, after the war he turned on others. He was rather disagreeable to many in the commune. Zadkine and his wife, they of course never forgot what this person had done."

"The war destroyed his family," Blanche said simply. "And they disappeared."

"No, Maman," Simone gently corrected her mother. "They lost the land, certainly, but some of them survived and still live in the area." What is certain is that Monsieur Verny eventually had to sell his property together with the once-magnificent Château Ladoux, and that many in the village ostracized them. The next owner wasn't interested in farming, and thus one more blow fell on a village already reeling from the loss of its sons and the post-war economic depression as one of its largest and richest properties, one that had employed a significant number of local workers, went into decline. By the late 1950s, Les Arques was in full retreat from life.

One of Zadkine's friends, a Belgian named Marc Edo Tralbaut, wrote of the village:

> Les Arques is one of those villages abandoned to its sad fate. It is not entirely deserted, but nearly so . . . Weeds grow up through the dirt of ground floors of the houses, while the stories above are open to the rain, hail, and snow, the roofs no longer repaired after the damage of storms. What is left of the doors and windows, a few boards barely in place, clacks in the wind. Here and there a ruin has been turned into a chicken coop or a place to store a few boards . . . One curious thing, in the middle of all of these buildings slowly dying and which will never come back to life, there is one which has been reborn, like a Phoenix, from the ashes. This is the ancient romanesque church . . . [Quoted in P. Béghain's Writers and Artists of Quercy, © 1999 Editions du Rouergue.]

Blanche Delfort put it far more simply. "When we came, there were people everywhere. Then the old people died, there was the war, the young left, and *voila*!"

As if Zadkine had not done enough for the village, it is also on his account that the Eglise St. Laurent was restored to the pristine, (very) approximately original, state the visitor finds it in today.

The crypt was dug out, the intricate interior stonework exposed and restored, and the hideous Gothic belltower, so out of place atop a Romanesque church, pulled down and replaced with a far simpler and more harmonious structure.

Zadkine in the fifties was nearing the apogee of his reputation as France's greatest living sculptor, winning awards and commissions, and a man of influence in the arts. It was his lobbying that got the church classed as a *monument historique* in 1952, the French equivalent of placing it on the National Register of Historic Places. The building was analyzed, stabilized, and, after further appeals to André Malraux, de Gaulle's minister of cultural affairs, restored completely in 1959, the first step on the long road back to life for the village. The next step would have to wait for the maturing of another generation, for the bitter memories of the war to pass into the grave with those who had lived them.

As for the true significance of Zadkine in the ongoing drama of this village's struggle to survive, I think it is best summed up by another mundane but telling comment from the museum guest book: "The welcome, the museum, the artist, the village, the ringing of the church bells and everything I am forgetting—thanks for all of it!" This visitor has understood that the man and his art are but one attraction, one which tends to get a lot of attention but whose importance, without the landscape, the presence of the church, and the people who have brought all of this alive, would be greatly diminished.

FIVE

A LIVE CHEF

My philosophy of cooking is, above all, about giving; not without limits, but generously. I don't think you succeed by saying, oh, this red mullet is so expensive I'm going to use only six ounces per serving instead of eight. What's good to do is to be able to use products which are expensive. You economize on one thing to give back on another. When truffles are in season, cook with truffles. And when I say cook with truffles, I mean so the customer can see, smell, and taste them, a couple of real generous slices. Cooking with lobster means not just a claw, or a tail, but a whole lobster with both claws, beautifully presented out of the shell. People feel that. They know it. Generosity makes itself felt.

— JACQUES RATIER, CHEF/OWNER OF
LA RÉCRÉATION IN LES ARQUES

While Zadkine came across the ad for his Les Arques house in a Parisian magazine, Jacques and Noëlle Ratier happened to have been watching the news in their Toulouse apartment one night when a very interesting tidbit crossed the screen. There was bashful Raymond Laval in front of the derelict Les Arques schoolhouse, which had been turned into a short-lived, now derelict, restaurant, looking for a chef willing to gamble that he or she could make a go of it in the hinterlands of the Lot. The year was 1992, and both Jacques and the village were at a crossroads.

Although the restoration of the church had been completed more than two decades earlier, the houses of the village were slowly decomposing around it as their owners died and the young moved away, uninterested or unable to keep the family home in repair. The school, the last vestige of life, closed its doors on the final class of twelve in 1986. The last commerce, Bousquet's store, which, like Blanche Delfort's earlier, had sold dry goods and served as a bread depot and even dispensed the occasional drink in one corner serving as a café, closed with it.

With the population in steep decline, the school newly empty, and the political will to make even the most minimal investments in roads, public sewers, and streetlights lacking in the departmental capital of Cahors, it appeared that Les Arques, like a thousand villages in rural France, was not so much dying as slowly, inexorably crumbling away to nothing.

It was at this dark moment in 1988 that Mayor Laval and his son, Gérard, and a handful of others involved in the arts put together a proposal and secured the funding to initiate a summer gathering of visiting artists; it was the birth of Les Ateliers des Arques. That first summer, the artists worked by day and gathered in the school at night to give talks, exchange ideas, meet the public, and carouse. There the restaurant was born, named La Récréation to reflect both its schoolhouse venue and the fact that going there was "recess," a break, for the artists.

Patrice Béghain, at the time the quite powerful DRAC, or regional director of cultural affairs (for the largest political region in France and with a budget to match), was largely responsible for initiating and overseeing the state's investment in the museum and Les Ateliers. Though he first came on business, he, too, fell in love with the village and eventually bought a house there. "At the same time as the Ateliers was coming into existence," he pointed out, "the commune decided to reopen the school as a restaurant. It was under many different managements [at first, but then] it was the commune that ran it. That place, the restaurant La Récréation, played a very important role. It was where the artists could get together to have a drink or eat a meal, a central place and important because they couldn't cook in the houses where they were staying. I went there often at the time to have a glass of wine, to eat. There were some memorable evenings at La Récréation where maybe we drank a little too much *vin de Cahors* or old prune eau-de-vie, but where ideas were exchanged, opinions aired, and where the novice could sup with the old hand at the same table."

The first year, the commune leased out the whole operation to a chef, who opened in the summer—and closed in the fall. A second chef appeared, but he, too, quickly folded his tent, unable to meet the overwhelming challenges of making a restaurant work—and having a life—in such a small, out-of-the-way village. Then the villagers ran it themselves with a hired chef and everyone else pitching in. Suzanne Laval, for example, washed dishes, and Patrice's daughter, Véronique, waited tables. "It was a disaster!" Raymond recalled, shaking his head and frowning. "Not professional, and we couldn't make any money, certainly not enough to keep it open." (Apparently they had a problem collecting tabs run up over the summer, and not just those of the itinerant artists.) Thus it went until, eight years ago, empty once again, with a leaky roof and crumbling interior plaster, vermin-infested and with rotten floors in the dining room, the building reached its nadir.

Today, it has become something quite different. From the outside, you see two chestnut trees shading a court of beige gravel in front of an imposing two-story building of local stone. The courtyard is framed on either side by the long, low open-fronted sheds where people used to tie up their horses. The front of the court is bordered by a low stone wall capped with a cast iron fence and, in the center, six-foot-tall gates painted that deep maroon of an old Jag. Between the chestnut trees strings of fairy lights float, and four large, square umbrellas shade the tables and chairs of the diners from the fierce midday sun.

Everywhere, as in the village, there are flowers, shrubs, trees blooming. Pots of geraniums, impatiens, begonias, nicotiana purple and white and pink, and yellow gardenia clutter every flat surface, and in a small chef's garden in front of the kitchen windows a bay laurel tree, mint, thyme, sage, and other herbs flourish. Bamboo soars twenty feet in one corner, butterfly trees in another, with beds of black-eyed Susans, flame, and purple globeflowers at the base of the wall in between.

A small protruding terrace, its tables and chairs almost completely hidden by a profusely blooming fifty-year-old wisteria, marks the entrance. To either side, the walls are pierced by eight-foot-tall, narrow windows in sets of three, their trim of robin's-egg blue. To the left of the terrace is the kitchen, and from outside you can glimpse the blur of chefs' whites, or, taking two steps in the front door and looking left, peek right into the kitchen at the prep work going on. To the right is the dining room, its windows lace-curtained. The layout of the restaurant is very simple, two rooms each about thirty feet square separated by a smaller room between, the back wall of which is taken up by an impressive fruitwood bar.

In the *salle*, the dining room, marigold tablecloths with flower blossoms, plush Chinese-red cushioned wooden chairs and walls the color of old linen invite you to relax without calling to mind how much they must have paid the decorator. Against one wall is a massive antique chestnut sideboard, its worn and much-scarred top covered with bottles of red wine. On the other side of the room, a blackboard with a chalked menu hangs underneath a high shelf displaying a set of ancient copper cookware. Noëlle uses the walls to feature the works of local artists, an ever-changing selection tending toward the colorful and the playful rather than dark abstraction.

Although the village pitched in for some of the first, major repairs (replacing the decrepit roof and rotting dining room floor), what you see today represents eight years of Jacques and Noëlle's labor, and of the restaurant's profits. They have constantly upgraded the building, often at the expense of their own needs. In the cave, a cavernous new walk-in freezer stands at one end, and an equally gleaming (and expensive) water-softening system at the other. Upstairs, in their apartment, he has started a total renovation, replacing a kitchen and bath and doing over the large bedroom, for he turns his hand to electricity, plumbing, and masonry work as well as carpentry.

"It doesn't look anything like how it did when we moved in," Jacques told me. "All the doors and windows were rotten, falling out of their frames. The roof was in terrible shape. When we had a hailstorm, the hail would get blown in under the clay tiles, melt, and run down the inside of the walls of the dining room!"

"We were just about to give up on it," the mayor told me of those darker days almost ten years earlier, "because it had turned into a very expensive lesson for us. The chefs, they could never pay the rent, and couldn't seem to make it work outside of the summer. People were unhappy that it was costing the village so much money, and yet it was always failing."

It was at this point that Mayor Laval had taken the desperate measure of running the commercial Jacques and Noëlle had seen. The offer was particularly enticing because it appeared a reasonable proposition: the commune promised a long lease at a low rent, an eventual option to purchase, and help with the rehabilitation to a qualified chef who would agree to live in the area and to keep the restaurant open at least seven months of the year. Mayor Laval received more than three hundred applications from all over the country.

"When we responded to the ad," Noëlle told me one day, "it was completely without reflection!" We were at the single marble-topped table that occupies one corner of the restaurant's bar, the small room separating the kitchen from the dining room. Jacques and Noëlle sat in the high-backed banquette of brown leather against the wall across from me, Nougat, their yellow Lab, snuggling between them, moaning from time to time as Jacques absently stroked his head.

It was late in the afternoon of a weekend toward the end of

fall, and the last of the supper crowd was trickling out in dribs and drabs, always stopping to salute the chef, customers in that happy haze resulting from a good meal and a few glasses of wine delivering a small, always positive, comment on some aspect of the food. Jacques was unfailingly polite, if a bit distant. He doesn't really listen to compliments, he confessed one day, nor set much store by them. He told me once he wished they'd be a little more critical.

Fred, Jacques' right-hand man, lounged in the doorway of the kitchen, smoking a Marlboro and listening to stories he'd never heard before. Jeanette and Sandrine came and went, shuttling trays of dirty dishes and glasses to the dishwasher, pausing to fire up the expresso machine for the few lingering customers. Everyone was moving slowly, an unhurried finish, for they serve only the one big meal on Sunday, and the season, too, was winding down, their winter hiatus just weeks away. The day was fresh, gusts of wind chasing fallen leaves around the courtyard, the sun streaming down with just enough force to recall summer's heat, and Jacques had opened the door to the terrace.

Like so many of our conversations, this one was spontaneous, for I had just wandered up to the village to see if anything was going on and happened upon them in a relaxed moment. As always, Jacques had asked me if I was hungry, insisting it would be no trouble to scare something up. I refused the offer of food, knowing it would delay the staff's departure, and accepted instead a glass of the sparkling Vouvray they were drinking.

"The first time we drove by and saw this place—" Jacques stopped and drew in his breath sharply. "We stood outside and just looked at it. I said, '*C'est ça qu'il nous faut!*'" We have to have it. "It was in February, cold, and there was snow covering everything, leaves everywhere, so sad-looking, abandoned. I looked at the kitchen inside. Everything had been redone, all the tiling. It didn't have the equipment it does now, but lots of light. We *had* to have it."

Jacques was then almost thirty years old, and he had spent half of his life in the kitchen and was having serious doubts about spending the rest of his life there.

His fascination with the kitchen had begun when he was quite young. When he was a kid, he explained, "all the other thirteen-year-old boys I knew were out running around in the street. I was at home, in the kitchen, learning how to make tarts and pastry from my grandmother, and learning how to cook from my grandfather. He was not a chef but he was a fantastic cook. Italian. He made things that were in fact very simple, but which were unbelievably good—rabbit in tomato sauce, roasted chicken, his own pasta and sauces. He was from a culture of using only the best in the kitchen. From May to October he ate his own vegetables and fruits. Never would he buy ham at the supermarket. He had it sent from Italy, it was the best, and when it was gone, you ate something else. That's the kind of education I had, the culture of growing up *à la campagne. C'est bien important!*"

From age sixteen, when he had taken fifth place in the annual Best Apprentice Chef in France competition, his career had moved on a steady upward arc. His showing guaranteed him a very good job, and he went straight into André Daguin's kitchen on the lowest rung, as a *commis* at the Michelin two-star L'Hotel de France in Auch. Several years later he moved on to the three-star Le Moulin à Mougins, under Roger Vergé. He stayed four years, leaving for a year in the middle to do his compulsory military service. ("My official and very important function was stretcher bearer, but actually I was cooking lunch and dinner for two hundred and fifty sailors on a Navy corvette.") By the time he left Le Moulin, he had risen to *chef de partie*, the number three spot in a French kitchen, responsible for all the sauces in Vergé's venerable establishment on the Côte d'Azur. (The ranks in a French kitchen are, from top to bottom, chef, *sous-chef* or *second*, *chef de partie*, *sous-chef de partie*, first *commis*, *commis*, and apprentice.)

After all those years of low-pay, high-pressure kitchens, and lured by fat offers and the chance to escape the provinces, he went AWOL, taking a series of jobs on a four-year tour through the best restaurants of the French West Indies. He acquired expertise in a number of different cuisines while also learning to scuba dive and windsurf, and spent the money as fast as he earned it.

"It's not the normal career path that I should have followed," he said with no hint of regret. "I didn't know *merde* at twenty-four, I knew only recipes. If I'd continued on that path a little further, maybe I'd be the *chef de cuisine* at the head of a whole brigade in a three-star—bossing a maître d'hotel, a sommelier, ten chefs in the kitchen, all that. It would have been too easy.

"I never gave a thought to my future," Jacques said of himself in his early twenties. "It was more a whole philosophy of how you should live. I was looking for sun, easy money, and women. I first started to make real money—five thousand dollars a month—in St. Bart's in the French West Indies at Le Sapotillier. In the next years, I worked at two more well-known restaurants in the islands, Le Restaurant du Port and Alizéa, both in St. Martin. I already knew I was a good cook. And at the Sapotillier, it was more like a performance, too, because you cooked right in front of the customers."

When Jacques talks of his time in the islands and cooking on luxury sail cruisers, his eyes light up and the words come fast. Two things seem to have happened during this time of happy exile. He found his own style in the kitchen, and he gained that rock-solid confidence in his own abilities which every truly accomplished chef seems to have. "I was always learning from others and from my own experience. How to use all the local tropical foods and fish, how to organize the kitchen better. The ship goes one week here, next week there, and the local products change. What you prepare changes with them."

It was at Le Sapotillier, where he had his first job as head chef, that he began to understand his own talents more fully and to see the first glimmers of where they might one day lead. "It was simple French food—for a one hundred percent American clientele—but it looks complicated, throwing together a duck salad, sauces flying around. It was a show. And that made me realize, wow, I can work anywhere. I did it well, and it was coming from my own creativity, my taste. [Today] if I make something for you, I'll tell you, 'You're going to love this.' And I know that because I made it, and I wouldn't serve it if it wasn't just fantastic, the best ingredients, well-seasoned, a good balance of flavors. That kind of thing, the ability to create a taste, it's innate, I think. Maybe." He cocked his head and, smiling, added, "I say that without pretension."

"He doesn't lack self-confidence when it comes to his own cuisine," Noëlle added a bit dryly.

In the years just prior to discovering La Récréation, he had tried ever more different things—head chef on a luxury Windstar sailing cruise ship in Polynesia, private chef on the yacht of an Italian arms dealer, and then, back in France, doing short stints in kitchens in and around Toulouse.

"But I was bored, I was restless, and I was unhappy."

"Remember that place you walked out of, Jacques?" Noëlle asked.

"*Putain* but that was disgusting. In Toulouse," he explained, "many of the kitchens are underground, in the caves of the restaurant. In that place—I'm not even going to tell you the name—I was working the stove and I felt something drop on my shoulder. I turned my head and saw a big gob of grease. I looked up, and the underside of the ceiling was just dripping. I didn't even finish out the service; I just packed my knives and left."

By the end of 1992, he was so burned out that he had just about decided to become a carpenter when the opportunity to remake La Récréation came up. How did they prevail over the 299 other applicants? "Jacques made a big impression on the village council." Noëlle said. "He'd call up every week, 'Have you made a decision, yet, Monsieur le Maire?' "

"I have a very strong CV," Jacques added, shrugging. "Of course, we had no money. We told them we had one hundred thousand francs, but we only had six thousand in the bank [about nine hundred dollars]. My stepfather helped us at the start, and we were able to repay him very quickly. Everyone pitched in. We had no money to hire people. My mother worked in the kitchen, my stepfather, too. Two cousins waited tables and helped prep. It was like that."

In May 1993, when they opened, Noëlle recalled, "We were starting a love affair, we were starting a business, we moved, all at the same time. It wasn't easy for me to make the decision," Noëlle said. "No one was holding a knife to my throat, making me quit my job in Toulouse to go live in the country. I wasn't sure I could adapt, that things would work out with Jacques. So I went to my parents, asked their advice. My father said, '*NON!!!*' Maybe just to spite him, I said, '*OUI!!!*' "

It seems they made the right decision, for this energetic young couple ended up creating, through some combination of atmospheric and culinary alchemy, a thriving restaurant in an unlikely place. From June through September, visitors from other parts of France and Europe, particularly Holland, Germany, and England, have "discovered" it to the extent that there is rarely a free seat in the house.

The rest of the year, however, La Récré belongs to the locals. Over the years they have slowly transformed it into the village crossroads and meeting place, holding the same importance to Les Arques as the diner or post office does to any small town in America. If something is happening and it involves more than six people, it is fair to say that it is probably happening at La Récré. The Hunters' Supper, the Golden Agers' Banquet, wedding feasts, anniversaries, birthdays, and First Communion celebrations, official gatherings for the Zadkine Museum and every local government event involving a meal—all of these take place at the restaurant.

Not that this happened overnight, as Jacques and Noëlle are the first to point out. "When we started," Noëlle told me, "it was not at all like this." She remembers times during the first year when they had but two customers between Friday night and Sunday noon. "Making a restaurant succeed is not a war you win overnight," she said. "You win it meal by meal, every night. When we first began, there was this ancient retired farmer who used to sit out on a lawn chair and watch the cars pass. He lived just across the street, and each morning I'd come out and he'd say, 'Thirteen last night, that's not bad, my young friends, not bad.' He was trying to keep our spirits up. He died before we had really become established, and I always wonder what he'd say now, when the parking lot is always full."

"That was hard." Jacques added. "But my *boules* game improved dramatically. The thing of it was, everyone wanted it to work so badly, wanted for us to succeed. The villagers, those first years they would come by and order the smallest thing, a coffee, an ice cream, to support us. That's how it really took off, first with the village and then people from Cazals."

"And that first winter, ooh, it was so cold!" Noëlle shivered at the memory.

"If it hadn't been for Bernard and Jeannine, we probably would have left!" Jacques added. Bernard and Jeannine are the Bousquets, farmers who work a huge kitchen garden for Jacques but, more importantly, as the restaurant's next-door neighbors, have become good friends.

"We had no heat, only wood stoves. There were customers eating dinner in the dining room with their coats on," Noëlle continued. "And you had to carry the wood up those stairs [to their apartment above]. We both got sick, really sick, in bed with fever, and Bernard and Jeannine just appeared with the wood — and food. Didn't make a big deal of it, just helped."

"I knew I was going to like Bernard from the first day we opened." Jacques said. "May 8, 1993, was the opening of the restaurant. We invited the whole commune, the priest, the gendarmes, the village, everyone. We had fifty or sixty people come, and we made *un buffet à la campagnard*. Country ham, pâté, salads, and plenty of wine. Fifteen people reserved for the next day, and Bernard came up to me. 'Can you do the Hunters' Supper?' he asked me. 'For fifty people?' Well, how much were you thinking to spend?" I asked him. "One hundred francs a head? You bet! Just like that he gave us a chance, and we needed the business! Back then we were serving a six-course prix fixe menu, soup, entrée, fish course, meat course, cheese, and dessert for one hundred francs, making no money at all. For fifty francs you could get soup, an entrée, a main dish, and dessert, with wine included!"

Today the menu at La Récré has evolved far away from its origins as Jacques has introduced his clientele to his particular cuisine, especially the huge variety of fish offerings he learned to prepare outside France. "Our formula evolved," Noëlle observed. "At first, it was much more traditional, like what you see in most restaurants here, duck confit, cassoulet, entrecôte, sautéed fresh foie gras with sauce . . ."

"It was one of the few things we ever argued about, my stepfather and I." Jacques said. "He was the one who said the menu should be more . . . what people here would expect. That first summer, we did foie gras, smoked duck breast salad, salad of duck gizzards, like Noëlle said. But after the season ended, I said, next year I'll do what I know how to do. Fish especially, that was what made our reputation. The Lotois don't eat it because it's not in their habits to eat it. They ate only *brandade de morue* on Fridays [most French are nominally Catholic], made from dried stockfish distributed by peddlers coming down the Lot River."

"It has ended up quite differently," he continued, "and obviously worked very well. One of the last things he said to me, my stepfather, when he was dying in the hospital a few years ago, he told me that he had been wrong. He had been in the restaurant supply business, and he had helped us so much, so much when we began, but he wanted to tell me that, to trust myself."

The Lot, in fact the whole southwest of France, is littered with restaurants whose menus are splashed with the words "Quercynois," "Périgordin," or *à l'ancienne*—all meaning traditional, of the region. They tend to feature the acme of the bourgeois cooking of southwest France, what you would have been served at the mayor's house for Sunday supper in the time of Proust—heavy goose, duck, and pork dishes with sauce, or in pastry, often truffled and morel-ed, foie gras-ed, and saffronned to boot. The result was immensely rich—and hard on the liver, as they say here.

Down the road from Les Arques about twenty minutes, in the village of St. Médard, for example, the gourmand with a taste for Michelin stars can find Le Gindreau, a one-star *restaurant gastronomique* run by chef Alexis Pélissou, a man famous among the locals in these parts for his impressive waxed mustachios as well as for his food. On his menu, one finds among the first courses wild porcini mushrooms in garlic cream sauce with croutons, a conserved foie gras served with a sauce of grapes sautéed in butter, escalopes of fresh foie gras with sour fruit and spices, langoustine ravioli in ginger butter, saffronned green vegetables in puff pastry, and scrambled eggs with truffles. (That last, by the way, will lighten your wallet by thirty dollars.) And this is before soldiering on to the main dishes.

"We used to have three prix fixe and one à la carte menu, we used the traditional local products, and," Noëlle added, laughing, "we didn't know our clientele at all!" These days, for lunch they also offer a single *plat du jour* or daily special plate with wine and a dessert for about fifteen dollars and a five-course prix fixe menu for about twenty-five dollars.

The prix fixe begins with a soup, moves on to a choice of five or six first courses, and the same of main dishes and dessert. In between these last two comes the *cabécou*, a disk of snowy-white, tangy goat's milk cheese made by Françoise Reix just up the road in Lherm from the milk of her own goats. With a carafe of local *vin de Cahors*, the sturdy local red, and coffee included, most people seem to find it an incredible bargain.

Although Pélissou would deny it, probably pointing out that it is impossible to compare his Michelin one-star and all that connotes with Jacques' more humble service, they do compete, if mostly for the tourist wallet. A reasonably watered five-course meal at Pélissou's establishment, with its sterling silver, crystal wineglasses, tuxedoed waiters, and lush decor, costs three, if not four, times as much as at Jacques'. In my time at La Récré, I often heard a customer at the end of one meal making reservations for another, often in the same week. Summer vacationers and residents here for a few weeks can and do come to eat at La Récré several times, while stopping at Le Gindreau only once.

They are genial competitors, however, for Pélissou has in the past asked Jacques to provide the blow-out end-of-season thank-you meal for Le Gindreau's large staff, a request regarded as an honor among chefs. Jacques and Noëlle favor Le Gindreau with their custom as well, and Jacques is not stingy with his compliments: "You can eat very, very well at Le Gindreau, and the food, the service, it is all at a very high level. And I admire him, too, because he went before us, started a restaurant in a schoolhouse with nothing in the middle of nowhere and made it a huge success. They come to eat here, he and his wife, and we exchange stories about our early days. What did you do when you ran out of bread in the middle of service? Borrowed from the neighbors? So did we!"

Jacques' menu has its own interior logic, serving both his needs and the customers' tastes. Take the soup. There's only ever one, usually something quite simple that changes with the seasons: gazpacho in summer, leek and potato or pumpkin or mushroom velouté in the fall and winter, and asparagus soup in the spring.

"We added a soup," said Noëlle, "because people in this region traditionally start their meal with a soup. For us, it's also a little trick which buys us time to prepare their entrées. They perceive it as an extra and always take it since it's offered as part of the prix fixe. And Jacques' soups are quite well known around here. I've heard people say to friends, 'Ah! You've got to try the gazpacho,' which we serve cold in the summer. They love his pumpkin soup in the fall, too. Pumpkins don't actually have much taste," she remarked, shrugging. The secret to its popularity? Noëlle smiled michieviously. "Butter. Lots of butter."

One afternoon, I had done an informal survey of several middle-aged couples taking coffee after lunch in the restaurant's court-yard. The couples, all English who owned second homes in the region and, it turned out, ate there quite often, mentioned how unusual it was to find this kind of sophisticated cuisine out here, so far from a city. They also remarked on how light Jacques' cooking is.

This is certainly true in that his food does not arrive at table swimming in sauce and that he doesn't put fat with fat (bacon-wrapped foie gras or sautéed scallops in cream-based shrimp/lobster bisque, for instance). But his *plat du jour* is much more than just meat and veg with a starch, and this is what the English couples were talking about.

That day, the *plat du jour* was salmon *aiguillettes*, two small wing-shaped fillets of meltingly tender fish on a bed of *escabèche*, a sauce made by cooking down shallot, carrot, vinegar, white wine, broth, and a bit of lemon juice until the liquid is much reduced and the flavor intensified. Rounding out the plate were a grilled tomato half, a mound of *cocos*, white beans with bits of prosciutto and parsley, and a squash blossom stuffed with mushroom mousse. Dessert, coffee, and wine are included, all for less than fifteen dollars. It is not the kind of meal that, either in quantity or richness, leaves you overful or reeling from the butter intake, and yet every little thing is so finely done, such an intricate balance of flavor, color, and texture, that you come away realizing you've eaten something very special indeed.

He does use lots of ultra-fresh, high quality if less expensive kinds of fish like *sandre* (a perchlike freshwater fish), sand dab, monkfish, even swordfish and tuna, both cheaper because they are not as well known here. It is almost possible (skipping dessert, alas!) to eat what, even to a diet-conscious American, is a light meal here. "Light," however, is a relative term in France; Jacques' kitchen consumes, in an average summer week, fifty pounds of butter and almost fifty quarts of cream. And this is French butter and French cream, both of which are significantly richer than their American supermarket equivalents.

"And during the summer rush," Noëlle explained, "you'll find changes in the menu two or three times a week because Jacques goes to market at four o'clock in the morning in Toulouse that often. We came up with this kind of menu to solve the problem of how to run this restaurant efficiently year-round, throughout those seasons when we're not always full. Then, you still want to keep the fresh things moving out the door. It's the kind of menu, too, which pleases the customers enormously because they can come in Monday and again on Wednesday and never eat the same thing twice. And when Jacques goes to market, if he sees some wonderful tuna, some incredibly fresh swordfish, he says, 'I'll take it!' And only then do we add it to the menu. So the menu's not rigid; it's very malleable."

Of the entrées offered, three almost never change: lobster ravioli in coral sauce, a walnut oil vinaigrette–dressed green salad topped with little packets of fresh goat cheese melted inside a fine pastry wrapping, and grilled red mullet fillets over a mildly spicy pico de gallo. Why? The short answer is that they are favorites. "They would kill me if I stopped making the lobster ravioli!" Jacques joked one day. "And the fish because some people want something light before a heavier main dish. The goat cheese salad? Who knows? People seem to like it and always order it."

Among the entrées that come and go as the seasons change are grilled fresh anchovies over greens, tuna or salmon carpaccio marinated in lime with shaved Parmesan, smoked salmon salad, a salad of fresh white beans with chunks of lobster with cilantro and cream, a fricassée of spring vegetables, warm lobster salad with truffle butter, pearls of Quercy melon in muscat, medallion of monkfish over a sweet pepper coulis, *flétan* ceviche with mint guacamole, lobster and grapefruit salad, squash blossom stuffed with scallop mousse, scallops and shrimp sautéed in a light pastry shell over tomato confit, Cajun-spiced shrimp over pico de gallo, seared scallops in orange butter over wilted mâche, sautéed fresh porcini, and artichoke hearts simmered with broth and spices, among others.

The same principle goes for the main courses, with the beef *pavé* in a wine sauce, salmon or perch fillet on a bed of *escabèche* in fine *brik* pastry wrapper, and duckling breast in a honey/lemon sauce almost always available even if their sauces may vary with the season. Here the variety changes enormously and often, everything from a lobster gratin to swordfish or tuna steak, saddle of

lamb with thyme, ground lamb with vegetables in a roasted eggplant shell, thin rolls of chicken breast stuffed with fresh wild
chanterelle or death trumpet or girolle mushrooms, leg of venison
in a reduced venison and red currant sauce, roasted pork loin in
summer truffle sauce, lamb brochettes in garlic and thyme sauce,
perch fillet topped with smoked duck breast, roasted guinea hen in
ginger sauce, civet of duck thigh with baby vegetables . . .

What is fresh at the market is what Jacques buys and cooks,
although he also takes advantage of very local seasonal abundance.
The wild plum out back provides the fruit for a *clafoutis*, and the
death trumpet, girolles, and porcini mushrooms he finds during
tramps through the woods also end up on the table.

Every so often during my time there, one of Jacques' neighbors would show up at the door with something in hand. Bernard,
out hunting in the morning, would appear in the late afternoon
with a paper bag of fresh porcini, in spring a crate of late asparagus
good only for soup, or even, occasionally, a cut of venison or boar.
Robert, a tiny old-timer with a lovely sweet smile, walked in one
day and shyly propped a bag of his own fresh shelled walnuts on
the counter, each one cracked by hand with a wooden mallet. Even
Madame Carrié offered up her prize parsley bush to Jacques'
shears.

They did not do this for money, and would only reluctantly
accept what Jacques pressed on them, if they took anything at all.
It was really more of the pleasure in giving, especially something
special that they knew would end up as a fillip on someone else's
plate. Of course, they also know that what goes around comes
around, for Jacques does not forget his friends when they come to
eat.

Some of this generosity can also be explained in the genuine
pleasure the inhabitants of the village take in the presence of the
restaurant. There are many reasons for this. "In essence," Patrice
Béghain told me, "the reopening of the school, that was seen by
people, even if it had reopened as a restaurant and not a school, as
a strong symbol because in a village, the school always has a place

in people's hearts." In my own experience as well, I often heard those of the older generation (most of the village) use the phrase, "the restaurant in the schoolhouse" rather than refer to it by name. They have a very intimate identification with it stemming from the rosy memories of youth, for most of them had gone to school there, boys in the dining room, girls in the kitchen.

Another reason everyone is happy to have La Récré there is equally simple: in the place of an empty, run-down building, they have a beautifully kept up, bustling commerce that brings people to Les Arques and keeps them there, at least for a few hours. Once, I asked Élise Ségol, who lives across the street, if sometimes the noise, especially when people lingered late in the courtyard in summer, ever bothered her. "I *like* to see people coming here," she said, surprised by the question. "Young people, children, families, they come for the restaurant and the museum. In the winter [when the restaurant is closed and museum less frequented], it is so quiet, lonely."

In the year 2000, there were 6,370 paid entries to the Zadkine Museum while La Récréation served about ten thousand meals in the same period. These two statistics hint at a subtle tension that has always existed between the restaurant on one side, and on the other village institutions—the Zadkine Museum and Les Ateliers des Arques—that depend squarely on the public purse. In the beginning, up to and including the first few years of the tenure of Jacques and Noëlle, the fate of the restaurant depended very much on the village and its attempts, through the museum and the summer arts program, to keep up the forward momentum. At the same time, a franc spent on renovating the restaurant, of which the village was still the owner and landlord, was a franc not available to other projects.

There was also a more telling emotional aspect summed up admirably by Patrice Béghain. "Today La Récréation has become a restaurant like other restaurants, a very good restaurant to be sure, but one which is essentially turned toward the exterior. At the beginning that place, La Récréation, was turned very much, linked very much, toward the project of Les Ateliers des Arques although of course open to other people. It was linked to the opening of the museum as well because it was at the same time that the departmental authorities decided to renovate the museum... These three things—the museum, the Ateliers, and the restaurant—functioned independently but at the same time were connected in people's minds. And this project had all the right ingredients—gastronomy and cultural heritage. It's very French!"

Once the village council perceived that the restaurant had begun to make, if not a fortune, at least a living for Jacques and Noëlle, the natural question arose, "Why spend any more of our tax money on the place?" According to Jacques, the short answer was that the building was still falling to pieces around them. The mayor and council had promised to fix the roof and replace the windows and doors and disintegrating floors and to sell the restaurant to them at a reasonable price.

But the council had made only token repairs, and, seeing the value of the business the two had built up, seemed to be hesitant about selling, with the price very much in question.

A few months before their lease was up, when the situation had already deteriorated, Raymond and his councillors came down to explain what they were and were not willing to do. As Jacques explained it, it got very tense very quickly. "I was really nervous. Everything we had done, the risks we had taken, and that it might just not work out. They were looking around, pointing out this problem and that problem, and I could just see where they were going. If we didn't renew the lease, we weren't going to get our security deposit back. I got so pissed off! I told them I would sue them until the day I died. And then I told the councillors to just shut their faces, not another word, that this was between Raymond and me."

And he called their bluff. He closed the restaurant and began to sell off the kitchen. "Poor Raymond, he would come by and peer in the window and frown at the place where a piece of equip-ment had been. He was worried! They were worried." Jacques and Noëlle were worried, too, until Raymond appeared one day with a new lease—and, later, an agreement to sell them the restaurant at a reasonable price. This being the small village that it is, and its inhabitants eminently practical beings at heart, everyone involved seems to have moved on. Two years later, Noëlle herself became a councillor, and, on her resignation in 2000, Jacques was elected in her place.

SIX

FOIE GRAS: A TALE OF BLOOD, FEATHERS, AND GASTRONOMY

You never know what you're going to eat when a chef invites you to dinner. In the case of Jacques and Noëlle, who live in an apartment with only a kitchenette above La Récréation, the restaurant's big dining room is their only dining room, which leads to the odd but quite pleasant experience of sitting down to table in a restaurant in which you are the only guests.

The invitation came in the middle of October, the days of lingering light long gone. On our way to Cahors to shop the weekend before, we had noticed two very different and rather unusual phenomena. First, along the D911, the twisting major road that descends off the *causse* to Cahors from the north, macabre black silhouettes the height and shape of a man, their chests crossed with a bold X of reflective tape, had sprouted at seemingly random intervals along the roadside, thirty-four of them in total, just along the twenty-five kilometers between Catus and Cahors. The first I took to be a macabre joke, as Halloween was just around the corner. The second, a sudden proliferation of billboards advertising a single item, foie gras, made me scratch my head. The first thing I asked

Noëlle, upon arriving that night, was what their purpose was, these black figures haunting the roadsides. "They mark the place where a person was injured or killed in a car accident," she said casually, "to remind you to be careful. More people die the week of Toussaint [Halloween] than at any other time during the year in France. We French, you see, we tend to drive too fast on roads which are too narrow." Having ridden with Noëlle and Jacques, both of whom drive like they moonlight for Team Maserati, I had no trouble believing her.

Outside the restaurant, the evening air was chilly, fragrant with the smoke from farmers burning over their fields. Inside, it was warm and cozy, the radiators pinging, the tall windows creaking in a gusting wind. Amy and I, having taken a long afternoon trek with the dog, were both ravenous, a hunger sharpened by the lovely anticipation of having a very good meal. The season of cold soup and light fish had passed, and we were looking forward to something heartier, more in keeping with the robust peasant fare— duck and rabbit and game—that our neighbors were eating at home. After we had chatted over glasses of cold Vouvray and had consumed shallow bowls of rich pumpkin soup the color of saffron, Noëlle came out of the kitchen bearing plates empty but for wedge-shaped pieces of conserved foie gras and a basket of thin-sliced, toasted baguette. Though the description is equivalent to saying the Taj Mahal is a big, old building, foie gras is the enlarged liver of a duck or goose fattened and then force-fed grain, usually corn, for several weeks before slaughter.

I used to think of this delicacy as something found on the menus of very expensive French restaurants, something vaguely liver-y and, ahem, part of that group of organ meats that all of us have had the experience of discovering on our plates simultaneous with the thought, "I ordered *what*?" (A particularly upsetting first date with a *ris-de-veau*, aka sweetbreads, aka a calf's thymus gland, in Avignon at the age of fourteen comes to mind. I had thought it was veal with rice.) To the French, however, foie gras is perhaps the acme of farm cuisine, a traditional part of many families'

Christmas and New Year's meals and something worth waiting for the whole year. It is also very expensive, as much as forty-five dollars a pound prepared and direct from the farm, and it can be even more expensive in a big-city gourmet food store.

"This is Madame Maury's duck foie gras, and it's *tout simplement* the best in the area," Noëlle said as I took my first bite. "You don't know Madame Maury? She lives right up the road from you, past the mayor's house."

Though I hadn't known her by name, I did know who she was. Madame Maury must be in her seventies, a stout woman with fiercely black dyed hair and impressive dresses whom I had often seen walking, with a cane, along the side of the road. She is as well known at La Récréation for her appetite as for her foie gras, hewing to the old line when it comes to the traditional Sunday supper. "She takes *six* courses, which is how people of her generation grew up eating," Jacques had told me one day in the restaurant kitchen when I asked why he was preparing an off-the-menu plate of smoked salmon canapés. "It's for Madame Maury. This isn't even part of her meal, just an *amuse-bouche* to accompany the aperitif. She takes the soup, then an entrée, then a fish, then a meat, and then cheese and desserts. She always goes for the best, the special things, ordering Champagne to start, the lobster or fresh sautéed porcinis as a supplement, then a couple of good bottles of wine with the meal. She likes to eat!"

Sunday afternoon meals with the whole family gathered for a holiday or special celebration even thirty years ago used to go Madame Maury one further. You sat down at table in the early afternoon, usually after walking home from morning Mass, and didn't get up until dark, after consuming *seven* courses, with one roasted meat and one meat in sauce after the soup, entrée, and fish and before the cheese and dessert. Accompanying the repast would be an aperitif to stir the taste buds gently to life and at least three different kinds of wine paired with the entrée and main dishes, white for fish, red for red meat, a sweet wine or *vin moelleux* for the desserts. Those much-exercised taste buds would then be thoroughly anesthetized by a final glass of *digestif*, a pear or prune eau-de-vie in the Lot, before you toddled off for a nap.

"*Mais, c'est le top!*" Jacques said, leaning back in his chair with a look of bliss on his face and a plate empty of foie gras before him. Then the table fell silent as we all contemplated the obvious truth of that statement. Madame Maury's duck foie gras, with a light, airy consistency resembling that of fresh goat cheese, had a smooth, melting texture on the tongue, a taste intensely rich and nutty but not fatty and bearing no resemblance to any liver I had ever consumed before. As the diminutive slice on my plate grew ever smaller, I wondered if there were more out in the kitchen.

To be more precise, what we were eating was foie gras *entier mi-cuit conservé en bocal*, or whole foie gras half cooked and conserved in a two-pint glass canning jar. Seen still in the jar, floating in milky-yellow duck fat, its lobes folded over into one another and the top turned gray from contact with the air, it is not that impressive. Before serving it is gently removed from the jar, wiped of clinging fat, and the discolored portions sliced away to reveal a pink to beige flesh. The best has a uniformly dense, smooth texture, like fresh chèvre or perhaps butter five minutes from the refrigerator. This foie gras, unlike the commercially available variety in cans and jars, was only very briefly exposed to high heat, the flesh thus less likely to melt. Such a respectful treatment, although it means the result has to be refrigerated, preserves more of its flavor as less of its richness converts to simple duck fat.

This is in fact one measure of a good foie gras at the market. No respectable producer actually puts duck or goose fat in the jar along with the liver. What you see has come from the breakdown of the liver cells, which then release their fat, as the foie gras was preserved under heat and usually pressure. Too much fat means the bird was probably too big, its liver too big as well, too *fondant* or tending to melt, and thus of an inferior quality. And as for foie gras in cans, as any French granny will tell you, only an idiot, usually a foreign idiot, buys it, anyway, because who would pay so much money for something they can't even see?

All this talk of fat brings us to a brutal truth: foie gras is more than half apply-directly-to-arteries bad stuff. "But duck fat is not bad fat!" Noëlle insisted as my wife and I exchanged knowing smiles at the seemingly bottomless capacity of the French for self-deception when it comes to food. "There is scientific evidence that the people of the southwest, I think it was in the Gers where my parents live, they eat lots of duck and foie gras, and they have the lowest rate of heart disease in France. It's true! Scientific evidence! Studies!" she said, regarding our skeptical faces. True or not, although even the best foie gras may seem light at first bite, it is definitely something one is advised to consume slowly, rarely, and in moderation. Even here in the Lot where it is a regional specialty, a generous first-course serving, never more than two ounces, is about the size of two silver dollar pancakes.

Precisely because it is so ubiquitous on other restaurant menus hereabouts, Jacques does not serve it unless it is requested as part of a group meal. Even then, and this shows how common backyard Madame Maury–like operations are in these parts, one member of the group usually insists on providing the fresh foie gras so everyone knows it will be good. (Who would bring their own ingredients to a restaurant in America? And what American chef would thank them for the contribution to the meal, let alone allow the food to cross the threshold?)

One night, in our local bar, the Café de Paris in nearby Cazals, I asked the half-dozen people, a mixed group of young and old sipping espressos or having a glass of beer or muscat, how often they ate foie gras. Whenever possible, came the answer, which turned out to be much less often than they wanted to, usually once or twice a year around the holidays. How much did it cost, did they know? A simple question, but one that caused great consternation and rubbing of heads. Sandrine, Alexandra, and Gilles, all in their twenties, huddled together for a moment or two after the others admitted they didn't know. "We don't know, either!" said Alexandra, finally. "We have friends who raise ducks, and we either buy a whole duck or my parents get it as a gift around Christmas."

The others nodded. "It's just not one of those things you buy, or think about buying," said Sandrine. "It's too expensive."

This was a real puzzler because of that other seasonal oddity I had noticed, those billboards sprouting like mushrooms overnight, all advertising foie gras and its associated products. Perhaps this wouldn't have been so surprising had they not been almost the *only* billboards along the road, with the exception of those advertising the McDonald's restaurant in Cahors, whose signs one could find as far as seventy-five miles away, up in Sarlat in the Dordogne. (I guess those were for people who really had to have a burger, although fast-food traditionalists, if such exist, might be horrified to discover that it came topped, not with bacon or onions, but with smoked duck breast or fresh goat cheese.)

Ubiquitous but rarely consumed, expensive but without price, foie gras was certainly intriguing, and I set out to find a . . . what were they even called, the people who made foie gras? One of the first things I was to learn, from a farmer named Luc Borie, was that until even fifteen or twenty years ago, those people were most often called Grandma. Raising ducks and geese for foie gras was traditionally the province of women, a chore at once too menial for men but also thought to be better done by Grandma, who raised and fed her fowl by hand to accustom them to her touch before the violence of gavage, the act of force-feeding. Partly for these reasons, it is only in recent years that foie gras, wherever it exists, has become more than a strictly local business.

My first meeting with Luc, who quickly became an invaluable guide, was at the night market in Catus, a small village about seven miles from Les Arques across the ridge and nestled into the next valley. Crinkly-eyed and sunburned, with rough hands and a sturdy, well-muscled frame, Luc looks like the farmer he is, the most recent in a line going back more than six generations. He has brown hair cut short all round and gone missing in front, and round, gentle black eyes, and he speaks a French well-tinged with patois.

That night, dressed up in a short-sleeve blue-and-white-checked shirt and blue cotton work pants, he stood, relaxed and smiling, behind a towering stack of cans and glass jars of various sizes adorned with simple paper labels identifying their contents. These were the fruits of his labor, the *élevage* and gavage, or raising and force-feeding, of ducks on his 140 acres of land just outside town. I saw *gésiers* (gizzards), *rillettes* (pounded, potted duck meat), whole preserved foie gras, foie gras with truffles, *confit de cuisses de canard* (duck thighs cooked and preserved in their own fat), *manchons* (confit of drumsticks), confit with lentils, duck liver pâté, an even more rustic pâté called *fritons*, hare pâté with juniper berries, and enormous jars of cassoulet. Next to the pyramid, a cold case brimmed with fresh, vacuum-packed *magrets*, duck breasts, duck "*côtelettes,*" fresh duck sausage, smoked duck breast, free-range chicken, duck "ham," and, of course, foie gras *cru entier*, or fresh, uncooked foie gras, and foie gras *entier mi-cuit*, which also has to be refrigerated.

As he was busy with other customers, we talked briefly and made plans to meet. "You can stop by, come watch the gavage in the late afternoon," he said. "*Ça fait un peu folklore.*" It's a little bit hokey, he warned with a half-apologetic smile. "But then, you can come back after all the tourists leave, in late fall, and I'll show you the whole process, which is not at all 'folklore.' "

Now, truthfully, I expected that foie gras, like laws and sausage, was going to turn out to be one of things whose origins we are all better off not knowing too much about. In a word, gross.

But I had had fair experience of the everyday violence and seasonal brutality of the barnyard growing up. When I was about ten years old, my mother and then stepfather went through an ill-advised period of attempting to live a self-sufficient life—and dragging us kids along with them. We lived on eight acres of field and wood in rural Pennsylvania at the time, surrounded by dairy farms and cornfields. There was an eighteenth-century fieldstone barn out back, a stream ran along the bottom of the big field next to our house, and it was altogether an ideal setting for the experiment.

Though we did not, thank God, have enough land to raise cereal crops or have a dairy herd, in short order my parents had assembled a menagerie of barnyard beasts: a sow and piglets every year, a couple of steer for beef, two dozen chickens for eggs, and half an acre of garden. The first year we named all the animals, and learned about pig-proof fences, foxes in the henhouse, wild dogs, and animal diseases. In the fall when it was time for the beasts to go to Detweiler's, the butcher, it was hard to say good-bye. We stopped naming them after that.

Some years my parents would buy a couple dozen roasters and fryers, which we would fatten in pens in the front yard throughout the summer. When the days started getting shorter and the nights colder, the cauldron would go on the fire, out would come the tree-stump chopping block and the machete, and we would butcher them over a weekend, packing the freezer for the winter. I have vivid memories of Christopher, my youngest brother, at age four or five, twirling merrily around in circles, whirling a newly headless chicken, spouting blood from its neck, around his head. Our job was catching the birds, binding their feet, hanging them upside down until they passed out, then killing, bleeding, dunking, and plucking them. Mom cleaned them, and I can still remember the piles of unlaid, shell-less green-gray eggs, the smallest the size of tiny pearls, that came out of the hens.

One Christmas Eve, a particularly clever pig escaped the pen and got himself hit by a car. All the slaughterhouses were closed for the holidays, and what do you do with a dead, 350-pound pig on Christmas Eve? Someone was looking out for us. The driver of the car was a butcher, on his way to visit his wife in the hospital. My parents decided there was nothing to do but slaughter it, and the butcher agreed to pass by on his return to do the most important part, which was opening it up and removing all the offal so the meat wouldn't be spoiled. We had it hung up with chain from a downspout by his return, and it took him three cuts and less than a minute to dispose of his part of the job.

We built a fire in the center of the driveway, borrowed a huge kettle, and after that I remember a blurry night of struggling up and down the stairs with kettle after kettle of boiling water, of rubbing my hands raw trying to remove the bristles from the skin with stiff brushes. At some point the *Foxfire* book came out as a reference, and I believe totally inappropriate power tools were involved, too, as the night grew long. The next day was spent making sausage and scrapple, jointing the meat, and starting the hams curing with a mixture of salt, molasses, and saltpeter. I remember nothing at all about what I got for Christmas, but we did eat some rather strange cuts of pork for the next year.

While "delight" might not have been the apt adjective to describe my mental state as my date with Luc approached, I was nevertheless prepared for much that would have flipped the faux-fur crowd.

A few weeks later I turned off a narrow road and into a dirt lane lined on either side by stone houses and sheds and barns. It was late afternoon, but the sun shone down mercilessly, from an achingly blue cloudless sky. The first thing I noticed was that this was most definitely not a cleaned up showplace for tourists, but a working farm. There was cast-off farm equipment abandoned on the verge, untidy piles of feed sacking and construction waste in odd corners. The whole place had not the air of neglect, but of people too busy getting the day's jobs done to tidy up about the place.

I noted this also because I drove too far up the drive and was hit with a dizzying blast of air heavily freighted with a deadly, noxious vapor being blown out the back of a long low building. Driving past that shed, on a scorching day with the car windows open and not a breath of breeze otherwise, I had a good five seconds of head-lolling, eye-watering olfactory overload followed by a desperate search for a stick of gum, a cigarette butt, anything to inhale to change the scenery in the nasal region. Hell, an old gym sock would have been a godsend at that point.

Having grown up with farm animals, I know my way around the manure pile, but this reek was a new one to me. Cow dung, in the rural Chester County of my youth, was the country perfume, especially of an early winter's day, when the farmers were spreading it on their fields. It is sharply ammoniac, bracing, but not unpleasant. The pigpen had been aromatic, too, but pig waste stinks sweetly, a high-rotten smell more than anything. Chickens, first, it takes a hell of a lot of their droppings to build up to critical mass (by which time it makes topnotch fertilizer, too), and by then most of it has dried and compacted on the henhouse floor to the point where it is a minor irritation.

One further—goats, which I'd been around a lot here in France, as there are at last count three goat farmers and cheese makers within ten minutes of our house. The odd thing about goat pellets is that they have the same dry chalky aroma as fresh chèvre cheese, that distinctive smell that so distinguishes it from cow's milk cheeses. (While I was driving my daughter's baby-sitter home one night in our tiny Nissan Micra, my nose was tickled by an unusual but pleasant tang from the seat next to me. I didn't figure out what it was until we pulled up in front of her house, the bottom floor of which is taken up entirely by the family's chèvre-making operation. Aha!)

Duck shit, for I knew instinctively that this was the putrescence that had laid me low, is truly the Everest of excrements. As I parked the car, I wondered, with a little trepidation, what this gavage I had been hearing so much about actually entailed to produce, at the back end, this horrific result.

Once my eyes had cleared and I could take in the sights once again, I realized that the Borie spread was quite picturesque, in an eighteenth-century kind of way. Chickens scratched underfoot, and, beyond the weathered stone buildings with their clay tile roofs, there were cows with calves lowing gently as they gathered at the fence line, waiting for their dinner. On both sides as far as the eye could see, a checkerboard of fields of corn or pasture spread out. In the far distance, several miles away, I could make out, on the sunniest slopes, the orderly ranks of dark green vines of someone's family vineyard. Stunted oaks, their leaves sparse and barely green, grew on the poorest ground, bald patches of stone showing through between them. Unlike the land in the valleys around Les Arques, here even the bottomland, at the lowest point where water usually collects and runs, was more scrubby than verdant. It was very dry country, very poor land almost entirely of pebbly fields or stony pasture, and I could see why Luc had turned to ducks.

Yvette, Luc's mom, a charming woman wearing a smock over her brightly flowered dress, greeted me, taking me up past outbuildings and barns to a large open structure three stories tall, with a metal roof but no walls. Luc was up on a tall ladder, welding sections of feeder pipe whose purpose was to fill three twenty-foot-tall round aluminum silos with winter silage for all of the animals. There I met Luc's dad, Jacques, a copy of his son but shrunk down a size or two, with a mouthful of gold teeth and a strong, callused grip. Luc was in a hurry, he told me, because it was already late in the season *"pour faire la moisson,"* to harvest the barley and oats and corn in the fields, and his son was trying to put in place a new system to premix the grain, saving time each morning and night at feeding.

Jacques, now semiretired, wore faded blue work pants and a sweat stained blue-and-white-checked shirt, and wiped his forehead in the stifling heat with the French equivalent of a feed cap, more like a bicycle racer's visored *kepi*. Like his son, he spoke a slow country French, and we chatted as he handed the occasional tool to Luc and Jean-Louis, Luc's brother-in-law. After a while, when it became clear that Luc was going to be up there awhile, Jacques asked if I wouldn't like to see the farm. As we walked from barn to field and then up to where a flock of several hundred ducks grazed in a bare pasture, he talked about the changes he'd seen in his time as a farmer.

"My father died when I was young," he said. "He was always handicapped by a sickness, and so I had to make my way early. At first, I plowed with oxen! Yes, I am of the generation when agriculture passed from the Middle Ages to what you see now." He showed me a rusted single-bladed iron plow with its oak handles nearly rotted away lying forgotten in the corner of a field. It must have weighed a couple hundred pounds, the single blade that he had to guide against the push of the rocky soil thick, blunt, and looking as if it had come from a smithy's forge rather than a Massey-Ferguson catalog. Which it had.

Jacques pointed to an adjoining two-acre field. "That would take me two weeks to plow. Two weeks! I tried horses first, but the only horse I could afford was what others didn't want, the horse that tried to bite you, kick you. Horses weren't good because when the plow hit a rock—and there are rocks everywhere in the field here!—the harness bit into his shoulder and he would rear back and refuse to go forward! My God were we happy when we got our first tractor. That was in the sixties."

He continued his stories as we were joined by two other families who had come to watch the gavage. "That field over there?" He waved an arm to the other side of the road on which we were standing, a steeply pitched, dun-colored hillside crowned by rocks. "That field you might plow four times a year, and not once would your plow ever be set right, either digging too deeply and hitting rock or too lightly and just scratching the soil. That's because the land here in this part of the Lot, *c'est ni causse, ni Bouriane.*" It's neither the pebbly limestone of the ridge nor the clayey soil of the north. "You can have sand, clay, stones, and ferrous soil all in one field. How do you work that ground?" he asked rhetorically, shaking his head.

Luc, covered in dust, white paint in the back of his hair, in well-used shorts and a sweat-stained gray shirt that had long ago lost its sleeves, came up and introduced himself, then guided the small group farther up the road to the field where the ducks begin their journey into foie gras. There were a dozen geese gathered by a feeder, but he pointed out the hundreds of ducks grazing in a shadier part across the field.

"*L'image c'est l'oie, mais l'actualité c'est le canard.*" The image is the goose, but the reality is the duck, he began. Though this statement might have sounded rather pompous coming from someone else, it was said with such a gentle sincerity, in such a quiet but authoritative voice, that our reaction was to lean closer, for this man obviously had some things to say worth hearing. "We keep a few geese because, well, it makes a bit of folklore," he continued almost apologetically. "Grandma with her geese, you know, but we don't force-feed them to make foie gras.

"At the end of June," he went on, "we bought these ducklings, put them in a field of grain. This field here was sown with oats and wheat, and had grown up to here"—he put a hand down around his knees—"when they came. Now, you see, there's nothing left. They eat the grain, then the straw, then the grass down even to the roots. And after they're gone, the field becomes very green and lush from you know what. In two months, they'll consume about twenty-five pounds of grain each before we start the gavage, which takes two weeks."

"Why ducks and not geese?" someone asked.

"The duck we raise is a hybrid, which means they all tend to be of the same size, same weight, same physical characteristics. They're more resistant to heat and disease. They're easier to force-feed because you don't have to adjust the equipment or the amount of feed. And gavage stresses the beasts, which ducks tolerate better. Geese, you can have short ones and tall ones and all sorts of variations. Geese only lay seasonally, and when the days begin to get longer, they lay less. Which means, if you want goslings all year, you have to build a special shed where you can control the amount of daylight, and that also means ventilating it—more expensive! One duck will produce one hundred ducklings a year. One goose only about forty goslings a year. One male duck can cover—mate with—one hundred females. One male goose can cover only three females."

A chestnut mare with a foal wandered over, curious, then a donkey, who kicked out at the horse to try to get her to move away from the water trough at which she was drinking. Luc laughed. "Yes, we have a whole zoo here."

We ambled down to a cowshed, Luc explaining the basics of gavage along the way. "You only force-feed the male ducks. Females, their liver is only good for pâté. So they get killed.

"As for the goose, well, that's the popular image, isn't it, Grandma with her dozen geese, massaging her special mixture of grain down their necks. The goose, its neck is like a tube, straight from the mouth to the gizzard, while the duck has a *jabot* [a crop] in which they store food and which regulates the amount which goes to the gizzard to be ground up and then goes to the stomach. And then there's the personal, more subjective reason. Me, I prefer the taste of duck liver."

The gavage shed and what went on inside it were, for me, about as far from the genial image of Grandma tending her flock as I could imagine. To the French families I was with, however, it was a fun and educational side trip for Mom, Dad, and the kids, just another quaint country practice and, best of all, with a delicious snack at the end.

The first thing you notice is that the front of the shed is made of louvered material, with a discernible breeze created by the fans inside which are drawing huge amounts of air through it to cool the ducks. It was at the other end of the shed, where the air is evacuated, where I had been nearly overcome by the acrid stench of duck waste.

The shed is a single, roughly square, concrete-floored room divided into eight waist-high pens, four to a side with a central walkway running front to back in-between. Overhead is a track from which hangs the gavage machine. Jean-Louis was in a pen in front of us, which he had divided in half with a movable sheet of wood running almost from one wall to the other. He sat in the gap on a stool, taking unfed ducks from one side, holding them between his knees and straightening out their necks. The gavage machine, a large funnel with an overlong spout hanging from a swing arm above him, he moved over the duck, placing the spout between its beaks and holding them closed around it. Into the big funnel at the top he poured a measure of well-cooked corn from a bucket at his feet. When he pulled a trigger on the side of the funnel, it turned a threaded plunger inside the funnel forcing the corn down the duck's esophagus and then into its crop, which bulged visibly after the feeding was finished. The ducks weigh about ten pounds before gavage begins, and are fed about two pounds of cooked corn morning and night for two weeks, at which time they will have put on about another three pounds.

The feeding of each duck took about thirty seconds, Jean-Louis grabbing the bird, placing the funnel mouth into its beak, pulling the trigger, letting it stagger to the other side of the temporary wall, then moving on to the next. He handled the ducks gently, and to my surprise they didn't try to escape, or even to bite his hands. "They get used to it pretty quickly," Luc told me later, "but it does stress them."

Looking into the other pens, where the ducks had already been fed, I saw the canard equivalent of the first circle of hell, perhaps fifty ducks in a small space panting, their mouths open, their tongues hanging out. They lurched about, drunk with food. A low trough of water runs all along the wall at the back of each pen, and the ducks drank eagerly. The fans seemed to lower the temperature considerably and certainly kept the stench at a level two notches below overpowering, but on the whole it was not an image one wants to have in mind sitting down to a plate of foie gras.

As if to forestall those of queasy stomach, Luc moved into the background of foie gras, explaining that they do gavage humanely and as gently as possible. "If you hurt the duck, the liver isn't good for anything. It's got blood in it, the flesh doesn't taste as good. If you crowd them, or don't feed them properly before gavage, or don't make sure they're cool and comfortable, with ample space and water, you will never get a good foie gras." He also stressed that this is not industrial gavage, which does exist, some in France but mostly in Eastern Europe, ducks confined in boxes periodically fed with hoses. Luc's ducks are fed by hand, by men not machines. It is also important that the same person do the gavage every time because a familiar hand also apparently reduces the stress. Before the force-feeding, they're raised outside, in the open air, on one hectare of land per eight hundred ducks.

"The engorged liver that gavage creates is not a cirrhosis, nor a cancer," he insisted as we ambled down the hill toward the farm store. "It is an overfeeding," an extension of a naturally occurring phenomenon in ducks and geese: that they overeat and store fat in the cells of the liver before the long migratory flights during which they eat little or nothing. "It's a natural process," he emphasized, "because if you stop the gavage and put the duck back into the field, in four days their liver will have returned to its original size." Foie gras, he concluded comfortingly, is as old as the hills. "You can see representations of it in Egyptian hieroglyphs, and then the Romans took it from the Egyptians, feeding their ducks with figs and grapes."

After the lesson comes the reward, a tasting in the best French tradition. Luc's wife, Ghislaine, ushered us into the cave underneath their house, a barrel-vaulted stone ceiling overhead and more warm stone underfoot, shelves all around overflowing with displays of their wares. Dark-haired and dark-eyed, with strong arms and a voice to match, Ghislaine is no shrinking violet. She moves with the economy and assurance of the chef she used to be, at Le Gindreau up the road in St. Médard. She seems reserved, a bit shy, until you get her talking about her children, two boys and a girl from six to twelve years old, for whom she gave up her prestigious job. "You can't be a wife, have kids, and be a chef at a restaurant like that with those hours," she told me simply.

Ghislaine brought out jars of foie gras, pâté de foie gras, and fritons from a refrigerator and began to spread slices of fresh baguette with a thin layer of each. As she put a finished plate on the large table in the center of the room, a particularly famished visitor reached eagerly for a slice. "*Mais non, monsieur!*" You have to wait two minutes so it can warm up to the perfect temperature!" she scolded me in all seriousness. As we waited, Luc poured glasses of Ratafia and Château Fanon, both from Ghislaine's brother's vineyard. "Good honest wines," he said without elaboration. Neither was more than four dollars a bottle, and his description was apt, if overmodest.

As the samples rapidly diminished, I sought out Yvette, Luc's mom, who hovered, tiny and smiling, in one corner. She seemed to be happy at the line at the cash register, but regarded the rest of the proceedings with a faint puzzlement at the hoopla these people were making at what was, after all, to her no more than a basic of everyday life. I had never imagined there were so many things one could make from a duck, foie gras aside, I told her. Certainly in America we rarely saw such delicacies.

"Oh, no, monsieur!" she said. "Even ten years ago nobody here ate fresh duck breast, or dried it and sliced it like Bayonne ham, or made it into sausage. We didn't eat it any other way than *les cuisses et les ailes confites*!" She was referring to a very traditional farmhouse dish found today on every restaurant menu in the southwest boasting regional cuisine: the thighs and wings previously cooked and preserved in their own fat, which are then reheated to melt off the clinging fat and usually served with lentils or white beans. "Maybe, for the holidays, you bought *un cou farci*," she went on. "The neck stuffed with pâté de foie gras which surrounded a little nugget of pure foie gras. Like that, it was a surprise, a little treasure when you cut into it!"

This whole foie gras thing was obviously much more complicated—and interesting—than I had ever imagined, and I left that day with many questions unanswered and, best of all, an invitation to come back to watch the butchering and preparation of foie gras.

On a frigid morning at the end of October, I rose in the dark, dressed, and was out the door before six-thirty, plenty of time to get to the farm before seven, when the slaughter started. The villages of Montgesty and Catus were sleeping as I passed, the streetlights just going out as the sun rose. The dim light revealed a stark landscape of long shadows and fogged-in valleys, a whole palette of browns in the fields of stubble and dry cornstalks and the limbs of the first newly bare trees. Along the tops of the ridges, colorful splashes of red and yellow, rows of vines not yet having dropped their leaves, stood out starkly against this drab backdrop. The farm was utterly still when I arrived, the only sound the distant baying of a hound and a rooster crowing. I walked up the lane among the farm buildings, the faint, milky light on the stone and the mist-shrouded land all around making the years drop away, bringing back Luc's description of life as it been lived here 150 years ago.

"In the time of our grandparents and their grandparents," he had told me, "not so long ago, you would see flocks of ducks and geese in every farmyard, and for a very good reason. They raised and force-fed them and then slaughtered them, usually in the winter, and sold only the liver. The liver reimbursed them for the whole bird, which they ate, although that was not why they raised them. It was not a cash crop, because one hundred fifty years ago, there was no commerce on the farm or in the village. Very little. The only thing that you could sell for money was wine.

"Everything was made or grown on the farm to be consumed there. Every village had for example a *caravale*, a field held communally with the best soil and free of pebbles where they grew hemp, to weave into the cloth they made all of their clothes from. The hamlet right up the road there?" He pointed east to a small clump of houses, maybe twelve or fifteen on the next ridge. "Even that small a place would have a *tisserand* who wove the cloth and was paid with food or other goods. For bread they grew their own wheat. And they grew walnut trees, for the nuts.

"And of course they force-fed geese, mostly for the fat, rather than for the meat. *For the fat!* Not for using in preservation, because pork fat is better than goose for that but for cooking! And the walnut oil, they burned in little lamps, a shallow dish with a wick suspended above—you see them in all the antique shops now—*les calèmes*. They had walnut oil, back then, for lights. Oh, people make such a big cheese of the walnut oil now, eh? But it's not that good, it goes rancid fast, and back then it was used almost entirely for lighting! They had no petroleum yet, that was the next thing to come. So they burned walnut oil or candles.

"There's a village, Luzech-les-Oules, about twenty kilometers from here. *Les oules*—that's what they made in that village, an *oule* is an enormous rough-fired clay pot in which our ancestors used to store meat over the winter. They'd cook it in goose fat, and then pack it, heavily salted, in more fat, a layer of meat, a layer of fat, then bury the jars in the ground for the cold. Of course, it might be a little, er, strong by the spring. But that was before refrigeration . . ."

My imaginings came to an abrupt end as Luc popped his head out the back door to let the cat out and gave a start when I greeted him out of the gloom. He didn't offer me a coffee, only rubber boots and an apron, which I declined, to my later regret.

Half an hour later I watched as he hung the first three of 120 black and white ducks upside down in stainless-steel duck restraints. He grabbed one by the head, pulled it up until its foot-long neck was at a right angle to the body, and electrocuted it with a long jolt to the eye from a handheld device that looked as innocuous as a power screwdriver. He put a long knife into the back of its mouth and slit the veins under the tongue, then let the head fall back against the white wall, blood pouring in a scarlet ribbon out of its beak and into the white plastic pail underneath it while the tail flapped pathetically a few times in death spasm.

One by one he picked them up, still twitching and flopping, sending the occasional shower of drops across the room, splashing his face, his apron, my shirt and notebook with red telltales, and hung them on a rack over a vat of nearly boiling water, the whole machine brightly labeled "ROBO-TREMPEUR!"—a robo-dunker! (their exclamation point). The rack descended, dunking the birds for the European Community–approved time necessary to facilitate plucking. The small room was very hot, fetid, its white walls and concrete floor dotted with blood and clumps of feathers and feces, the occasional blasts of frigid morning air welcome, indeed, when Jean-Louis opened the front door to unload more crates of live ducks.

Next Luc manipulated the duck over two sets of rubber rollers with rotating nubbly spokes, which plucked the bird roughly, the scent of wet feathers and duck fear adding itself to the cloying steam rising up thickly from the buckets against the wall. Then two apprentices, from the agricultural lycée outside Cahors, went over the birds more finely with blowtorch and electric cauterizer and paper towels, burning and wiping until the pale yellow skin shone, the plump forms of their swollen livers pushing obscenely at their bellies, waiting for release.

The room was filled with all sorts of odd, unpleasant noises, the blat and roar of the blowtorch being lit and its flame adjusted, the dry *snick-snick* of Luc's long knife slitting veins, the brief flap and quack of ducks being taken from bright green and red cages to the death wall. No one spoke, aside from the occasional grunted direction from Luc or Ghislaine to the apprentices.

Dangling from a hook by one leg, the glistening bodies passed through a hole in the wall into the first of several cold rooms, where Ghislaine grabbed one off the hook and slid it into another rack bolted to the wall. This one holds the birds neck up, legs dangling down, exposing the belly. She cut into *les pattes*, the feet, at the first joint, snapping them off and dropping them into a bucket, then made a shallow U-shaped slit about three inches up from the tail. Putting the knife aside, she slid both hands up into the cavity and eased the foie gras, still steaming with body heat, out into the frigid air. As it emerged, a handful of undigested corn fell out of the cavity, hitting the concrete floor softly.

She looked the liver over, tsk-tsking. "That's a big duck!" She regarded it even more closely, cocking her head to one side. "It's too big," she continued, frowning. "Feel it." The liver, large as a dinner plate, she held flat spread over her two large hands. It was smooth to the touch, warm, its two lobes, the left larger than the right, reminding me of a pair of human lungs, only squished together. "The flesh is too firm, too hard." I pressed a finger into it, curious to understand what she meant, but felt only soft, minimally resistant flesh that reminded me of—liver.

She eased it gently into a basin of ice water on a trolley to her right, then took up the knife once again to slice away the dangling entrails where the liver had rested, letting them fall into another bucket at her feet. She reached in and grabbed the gizzard, which looks like an enormous blue mussel covered with a coarse yellow membrane, and threw it into another bucket. The partially butchered bird, looking like a supermarket chicken except for the long neck and head, then went into the cold room next door to await the *decoupage*, or final dressing into the appropriate pieces.

By the time she returned, three ducks had piled up at the entrance to her evisceration station. She racked the first, slit and cracked off the feet, and on it went, duck after duck, until the basin was full.

She wiped her bloody hands on a paper towel, then gestured for me to follow her next door, into another cold room, this one with a large stainless-steel prep table on one side, a white scale on its top. *"Alors, on va les trier,"* she said. Let's go grade them. She laid out three empty white plastic trays and immediately set to flinging the thirty or so livers from the big basin into one of the three trays according to their quality.

She picked one up, spreading the flesh out on the tabletop. The color was that of putty, uniform, the texture even and smooth, somewhere between a chicken breast and a nearly inflated football. She threw it on the scale. "Five hundred and fifty grams [one and a quarter pounds], that's about what we're looking for, the best weight. Too small and they begin to taste like chicken liver. Too big and it's the same, not good, the liver is too *fondant*, too tender such that it will lose too much fat, will disintegrate, when you conserve it under heat and pressure."

In a foie gras, the liver cells are swollen with fat. As soon as the bird dies, enzymes in the liver begin to digest the cell walls, releasing the fat, which is why they have to be handled very quickly, with no more than four hours passing between slaughter and preservation. Overly large livers mean the cells are packed with too much fat, the membranes weak and thus more likely to burst, releasing their fat with the slightest heat. Too small and there's not enough fat, and what you taste is more foie, liver cells, than gras, the fat.

"So now you see why we're in such a rush. You can't grade

them after they've been sitting. You have to butcher them right away and look at them then. You want them not too red and not too white in color. But if they stay in the cold water too long, they all get pale, the blood leaves them, which is why we only have a couple of hours to do them all, no matter how many there are to butcher. You can't kill all the ducks first, then continue from there."

"This is that first, huge one I showed you." She threw it on the scale. "Eight hundred and fifty grams, *mon Dieu*! That one's for pâté," she said, throwing it into the third tray. "This one," she continued, flicking with her fingernail a liver that had a green spot on it. "That's bile from the liver ducts which has passed into the liver itself. Not good. We'll cut that out, but still what's left is only a *tout venant*." *Tout venant*, ordinary grade, is the third and lowest quality, sold for about eleven dollars a pound and suitable only for pâté, or stuffing necks, or being mixed with other meats for canning. The top grade, extra, sells for a little more than twenty dollars a pound, the middle grade, premiere, for fifteen dollars a pound. These are prices for foie gras entier cru, or uncooked whole foie gras. Every level of processing adds value, thus a pound of best-quality conserved foie gras in a glass jar will cost closer to forty-five dollars locally, with the price increasing proportionally according to how far you are from the farm.

"These red splotches?" She pointed out a liver, one lobe of which was partially dotted with areas of what looked like red cross-hatching. "That's blood, from an injury. They mostly hurt their necks or their feet, the necks from the first gavage when they're upset, stressed. And the feet get hurt because they're together with other ducks in a pen, and can't move around as much. But they get used to the gavage after the first time."

After she had done triage on the first pan of livers, she began to put together the day's orders, for which she needed a good supply of the two top grades of livers and of duck carcasses. One hundred percent of their output is sold to individuals, people like you and me (well, perhaps not) who call up requesting raw foie gras to

conserve themselves, usually in glass jars with rubber seals prepared with the boiling water method exactly like home canning.
These people also usually take the rest of the bird, too, butchered,
as well as a couple of cans of duck fat, which the Bories always
throw in free even though it is a valuable commodity in its own
right. The duck fat is for cooking the thighs and drumsticks, making a confit that is then canned like the foie gras.

A whole duck comes not cut into breasts, thighs, drumsticks,
wings, etc., on a tray as in the supermarket, but in four parts, in a
large crate. First Ghislaine makes a cut just below the head, cracks
it sharply back, and puts it aside. She makes another cut encircling
the base of the foot-long neck but not cutting through it, then puts
her fingers under the edge of the cut, grabs the skin and strips it off
toward her as if she were removing a sock. Next, she holds the carcass neck up and bangs it sharply against the edge of the table two
or three times, jarring loose any remaining grains of corn from the
base of the throat. She turns the duck over breast side up and
makes a deep slit along one side of the breastbone from tail to
neck, then slices down along the ribs first on one side, then on the
other so that the two breasts come away still attached to the skin.
She continues slicing with the knife tip between skin and body
along the back again first on one side, then the other, so that the
thighs fall away, too, also still attached to the skin. Finally, she
turns the duck on end, slicing along the backbone from head to tail
so that the skin falls away in a single piece.

Very strong in the arms and hands, and adept with her stubby
boning knife, Ghislaine works quickly and silently, taking less than
a minute to reduce a whole bird to a pieced carcass, the only sounds
the crack of bones and *shush* of her knife through flesh. The carcass,
a rib cage encased in purplish scraps of meat, goes into the customer's crate, but not before she has pointed out a few things.

"People argue over what the finest morsel on the duck is," she
told me, smiling, obviously about to impart her own opinion.
With the carcass resting breastbone up, as if she were carving a
chicken, she slipped the tip of the knife under the thin flap of

flesh left underneath where she had earlier stripped off the breast. She worked the knife toward the tail and came away with a long, thin strip of flesh, part of what we call the tender, and which, on a chicken breast, is the long thin piece you cut away to make it flat. "That's the *aiguillette*, like the filet mignon on a beef. Very delicate." She touched the knife to a thin pocket of flesh still remaining just below the breastbone and in front of the tail, no more than a teaspoonful. "That's what we call around here *Le sot l'y laisse*," which translates as, 'Only an idiot leaves it there.'

She reached into a bucket at her feet and pulled out a pair of feet and a head, which she put into a plastic bag. "Before, people would eat the feet and head, cook them up with the bones. Grandmother would collect the tongues, cook them in fat, and keep them for when the grandchildren were sick, the tastiest morsel. People don't do that so much now, but still, we'd get customers who said, 'Where's the rest of it?' when we left them out." She shrugged. "So now, we give them the whole thing."

On the table is the *paletot*, which means jacket, looking like a coat spread out, at one end the almost purple meat of the thighs sticking up to the left and right with their protruding bones, at the other the roughly oval *magrets*, or breasts, and in between the skin covered by a quarter inch layer of fat. The *paletot*, too, goes into the customer's crate, as does one top-grade, five-to-six-hundred-gram foie gras in a plastic bag, and another bag containing the head, feet, and gizzard.

One order was for eleven foie gras and a couple of complete birds. When I observed that that seemed to be a very large order (it came to more than two hundred seventy-five dollars), Ghislaine shrugged. "This is a customer from Agen," a city about a hundred miles southwest. "His mother still lives here, and he takes orders from his friends for the holidays. Anyone older than fifty in these parts knows how to prepare foie gras," she added casually. "She'll cook and preserve it for him. He may not know how, or not have the space in the city. People order like this because they don't know how to butcher it or they don't have the time."

"Also, when we sell them the whole duck, they don't have to be all done up like in the market." She went to the cold room and emerged with a whole duck. She crossed the legs tightly, as if they were tied together with string, and held it up, legs down, displaying it as it would hang in the market. Crossing the legs pushes the midsection, where the liver is, toward the eye. "See, it looks like it has a big liver, but they're fooling the customer because there's a thick layer of fat there." She laid it out flat on the table again, and pointed out several pin feathers that hadn't been removed, a few other spots where the ends of feathers remained embedded in the flesh. "Our ducks aren't so perfect, so shined up, as in a market stall, but the customer sees what they get and they get the whole thing. That's why we cut up the duck for them when they order a whole bird. So they can see the foie gras and know it's a good one and not one like those with the blood or bile in them."

By nine o'clock, the occasional customer come to pick up an order had turned into a stream of people coming and going in everything from farm trucks to BMWs. Was it always this busy? I asked, remembering that foie gras, Luc had told me, was a largely seasonal business, despite the producers' best attempts to get people to eat it more than just for the holidays.

"Toussaint, the Day of the Dead [Halloween], this is the start of our season, now through March," Ghislaine replied. "Everyone who still keeps the family home here, they come back to bring flowers to the graves of their ancestors. Because if the tomb has no flowers, what will the neighbors say?" She chuckled, smiling a little wickedly. "They all go to the cemetery on Sunday, when the priest says the prayers among the graves. It's a week-long school holiday, too. So those who have gone away, they come back and they come here to order the foie gras for a big family meal."

Luc, having finished with the initial butchering, had been shuttling basins of fresh liver into the cold room as we spoke, and he had his own ideas about the delicacies of the duck. "Have you ever been to a *soiree os de canard*?" he asked me. A duck bone party? I thought.

Ghislaine rolled her eyes. "My husband plays rugby with a club on Sunday afternoons, and this weekend, after they play, they'll make an occasion of tasting the new *vin de Cahors*," she said. "It's not a night after which you want to have to blow into the balloon!"

"Our team is the Ratafios, after the wine my brother-in-law makes. Our bones aren't as hard as the kids anymore, so we go a bit easy, but we always make an evening of the new wine and duck bones. We'll roast chestnuts to eat with the new wine first, *after* we practice, of course. And you really shouldn't eat anything before, long before, let me tell you!" he added enthusiastically. "It's a bit of a boozer."

"On Friday," his wife continued, giving him a look that some-how combined indulgent affection with a touch of disapproval, "I'll take a dozen carcasses and heads and some wings, and I'll cook them up in duck fat for him."

"We'll start with a *tourin*," Luc said. This is a local soup made of onions lightly sautéed in goose or duck fat, then made into a bouillon with the addition of water. A big clay soup pot is pre-pared with alternating layers of stale bread sliced thin and grated Cantal and Gruyère cheese, the bouillon added, then the top cov-ered with cheese and the whole passed under the broiler until nicely brown. "Then"—he made a crunching sound with his mouth—"eat the meat off the bones with our fingers, then have a salad. It'll be a feast!" The effect was a little macabre, the two of them standing around in white boots and rubber aprons, Luc's bald forehead dotted with feathers and a streak of red gore where he'd apparently had an itch.

On the table, three tubs sat containing the rest of the livers from the 120 ducks they had slaughtered that morning. In what was obviously a well-practiced ritual, Luc took up a position in front of the scale with a pile of already-cleaned livers to his left. Ghislaine stood on the long side of the table, flipping livers over and cleaning away the mass of nerves and veins at their center. One of the apprentices stood at the far end, taking squat 250-gram (about eight-ounce) glass jars out of a box and fitting them with rubber seals and glass lids. They were portioning the livers left over after all the orders had been filled to make preserved foie gras. It was ten-thirty, and they still had a half hour before the four-hour deadline was up.

Working quickly, Luc eyeballed each liver and sliced away a lobe, then threw it on the scale to see how close he'd come to 180 grams. Then he threw on another small piece, maybe twenty grams, to bring the jar up to weight. "When you sell a two-hundred-gram jar of foie gras *entier*," he explained, "by law one piece in the jar must weigh about one hundred eighty grams." Ghislaine took a pinch of salt and pepper from a tub and dropped it in the center of the piece, folded it, and stuffed it into the jar. The apprentice attached the lid over the rubber seal with temporary clips.

All these jars were destined for the autoclave in the next room, where they would be immersed in water heated to 105° Celsius (about 230° Fahrenheit) as quickly as possible. The temperature is higher than boiling because the autoclave is sealed and thus under pressure as in a pressure cooker. In the autoclave, the heat makes the air in the jar expand, exiting under the rubber seal, which also keeps the surrounding very hot water out. When the jars cool, the volume of air inside the jar contracts, pulling down the lid even more firmly to create a tight seal. The heat kills any bad bacteria, but it also cooks the foie gras.

"We always end up cooking it more than we want to in order to follow the rules about temperature and sterilization," Luc told me later in the finishing room. The walls were stacked with crates of the various things they make, and I noticed a curious thing. Only the foie gras and pâté de foie gras were in glass jars. "That's so the customer can see what they're getting." Ghislaine explained. She pulled a jar off the table in one corner. "This is from the last batch we did. See how much fat there is, not just on top but around the sides? Maybe thirty percent fat this is, and we can't sell it. We didn't add any fat, of course; it's what came out of the liver when it was in the autoclave."

"You can buy foie gras in cans for more reasonable prices at some stores," Luc added. "But it's in a can, and you can't see it. There are some who put in a little foie gras and a lot of fat, so you're buying some pretty expensive duck fat!"

It was about eleven-thirty when they finished. In the course of the morning, they sold more than one thousand dollars of foie gras, the same of duck parts and carcasses, with approximately the same value in conserved products left over to be sold in the cave and at the two weekly markets, in Luzech and Catus, where they also sold their goods directly. Given that they had to purchase the ducklings, fatten them for six weeks in the field, then force-feed them for two weeks, it seemed like an awful lot of work for that return.

"We know we won't make our fortune," Ghislaine observed. "But if we make a decent living from it, that's fine." How about time off? Wasn't it hard to escape the farm when someone had to look after the animals? "We take a week off in June," she averred. "But we can't even leave at Christmas because we have customers coming and calling up until the thirty-first of December. This one wants two foie gras for New Year's, that one can't come until six o'clock on Sunday night. But we've known them for so long, for years, so it's just like that.

"It's a very taxing way to make a living, that's true. The whole idea of a thirty-five-hour workweek makes us laugh. So, yes, it's a lot of work, and complicated so that you have to manage your time. But it's a good life," she finished. "It pleases us."

Luc and Ghislaine's twenty years of farming in this hardscrabble corner of southwest France have been a roller-coaster ride of good years and bad, of turning their hands to first one kind of farming and then another in a constant battle with changing economic conditions, the government, the weather, and what the market demands.

Luc began after finishing agricultural college with a dairy herd, but soon ran up against the ages-old complaint of farmers here— too little land and not enough water. Without sufficient pasturage to put hay by, or cropland to grow the corn for winter, he was forced to buy ever-larger amounts of feed, which erased any profit from increased milk sales from improvements he had made to the herd.

"We had to find another route to diversification, and that's how we became the first in the region to force-feed geese and ducks for a cooperative, which bought the ducks ready for slaughter. It was something that I did in order to earn an income for *me*, for my own family. If we had continued to have the milk cows, it wouldn't have worked. So we sold the cows, we did well, and used all of that money to redo the buildings here for raising ducks and geese. We started that winter force-feeding four hundred fifty geese, then more with ducks in the summer. And at that time there was such a demand that we made good money—and profits like we haven't seen since! We were among the first to do the gavage in an organized, rational way. We bought electric gavage machines, and besides, we were the first *men* to do it. Because gavage was always women's work before."

But there was a problem. The only way to make a living was to expand the operation, not necessarily in numbers but in its scope. And yet, "it was not legal to dress the ducks on the farm at that point," Luc told me with a touch of exasperation. "You could raise, force-feed, and slaughter them, but then everything else had to happen in a municipal slaughterhouse."

"There was a man up at Degagnac [in the north of the Lot], a wonderful, incredible guy, *le père* Balmes everyone called him, a man who had won a National Medal for his foie gras even. And it's thanks to him that we were able to develop this business on the farm. At that time, there was almost no foie gras that was officially regulated made on the farm for sale in any store. It was just individuals who raised and slaughtered a few geese or ducks and then sold the foie gras to others at *les marchés au gras*—fattening markets. You can still see them today in the departments of Le Gers and Les Landes, where people go to buy freshly slaughtered fattened geese and ducks out in the open air of the market."

Before, there was a multitude of very small producers. There wasn't even, back then, any industrial production. "It was *le père* Balmes who proved to the minister of agriculture, besieged him really, until he was forced to admit that you could do gavage and slaughter and dress ducks on the farm to produce foie gras. And that fight had to go all the way to Brussels [before the European Community commissions on agricultural norms and regulations] before he won. His principal argument, his defense, was not only that it was possible, but that you could then ship it, and if you could ship it, export it. And so the EC actually helped us in the end, writing the regulations, developing the norms.

"To many," he continued scornfully, "the socialist utopia is when each has a part in the whole. And that's what has happened in the industrial production of foie gras. The gaveur who force-feeds the ducks, he is no longer the master of anything! It's the big feed companies who give him young ducks which someone else has already raised. He force-feeds them and then the company takes them back to process themselves or send to the processor. It's true he doesn't have much responsibility or risk, but his margins are like this!" Luc put thumb and forefinger together. "Almost nothing. So they have to force-feed huge numbers to make any money. We, on the other hand, we have eliminated all of the inter-mediaries, which permits us to live."

At the Borie farm, they raise about twenty-three hundred ducks a year. There is an industrial operation near Sarlat, in the Dordogne just to the north, which has applied for a license to do the same for half a million ducks a year. "There has been such a huge demand for these products at a low price that it has created a whole industry, which still can't satisfy the demand. Eighty percent of the foie gras sold in France is imported, mostly from Eastern Europe, Hungary, and Poland especially. In twenty years there has been this huge change. The industry has gone from really, well, not the Middle Ages but something similar, to big business."

Today, on their farm, Luc, Ghislaine, and Jean-Louis raise the young ducks to maturity, force-feed, slaughter, and dress them, then package and sell the finished product, all without the intervention of middlemen. Every aspect of their operation, and especially the slaughter, dressing, and conservation, is carried out accordingly to European Community norms. It has been an uphill struggle, and they work very hard indeed, just to make an adequate living. In addition to the foie gras operation, they also grow and package almost three acres of white asparagus in the spring, raise free-range chickens, and have a beef herd of more than one hundred animals.

They find themselves, as niche producers, at a very interesting moment, able to prosper because of the increasing value the French consumer puts on knowing as much as possible about what goes into his mouth. And if you can know the individual who makes that food personally, so much the better! At all the village markets, every product is labeled as to origin as a matter of course. It's not just mâche, but mâche from Nantes that is famous for this lowly salad green, strawberries from nearby Lot-et-Garonne, whose sunny, well-drained soils yield much of the local fruit and produce, at the cheese stall *le vrai camembert de Isigny-Ste-Mère AOC*, true, raw milk Camembert made in traditional wooden molds in Normandy.

The AOC, as in wine, means *appellation d'origine controlée*, in essence that the product is pedigreed, that it is certified to come from a certain area, to be made under strictly controlled and monitored conditions, and to contain just what it says it does. For this Camembert, the AOC designation means that the milk is not pasteurized, that it comes only from a particular breed of cow called *Normand*, grazed in a certain geographical area with limits on animals per hectare, that it contains a minimum percentage of milk fat, and even how long, in what manner, and where it is aged.

At the same time I was learning about foie gras with Luc and Ghislaine, a very serious drama involving beef and mad cow disease was playing itself out on the evening news and in all the newspapers. France was gripped by a panic stemming from a new and much more rigorous spot-checking program put into place by the veterinary inspectors. Inevitably, they found many more mad cows, including a few slipped into the meat stream by unscrupulous farmers whose animals ended up on the shelves of some of the largest supermarkets. As the fall progressed, beef consumption plummeted, butcher shops stopped selling many cuts, especially organ meats and those on the bone, school cafeterias across the country stopped serving it to schoolchildren, and poultry and fish sales soared.

For Luc and others like him, this wholesale cultural backlash against mass-produced, processed, or genetically manipulated food has given a new impetus to their small farms and to the sale of what they produce. Instead of watching their customers flee to the huge supermarket chains for lower prices, they have found that the products they produce artisanally and in smaller quantities are increasingly popular as consumers insist on doing business with people they know and trust.

No dewy-eyed romantic, Luc sees a void he's been planning to fill for some time. "We're going to start turning the whole beef operation into an organic farm," he told me. "First, there's a huge demand. Second, it simplifies our work enormously. We will convert about twenty acres a year to raise organic grain to feed the herd. I hope, in two or three years, to be slaughtering and butchering them here, too, and taking advantage of the clientele which we have already for the duck. After three years of converting to organic, it will be organic beef, certainly, but more importantly beef sold directly to the consumer from the farm. We cut out all the intermediaries, and so it will also be less expensive than organic meat they can find at the butcher. But it's really about image with organic food, the image of the farm that they're looking for and that therefore adds value to what we produce."

Can organic foie gras be far behind? "Well," Luc told me, laughing, "we're exploring raising our own ducks from eggs instead of buying the ducklings as we do now. And I'm interested in trying to raise the heirloom breeds, the more rustic Rouen and

Colvert ducks. They're less productive than the Barbary/Peking cross that all industrial foie gras comes from because they're easier to breed and to work with, but maybe the foie gras is better." He shrugs, a smile crossing his face. "Come back in a few years, maybe we'll have something new to show you."

SEVEN

DECEMBER

One Friday night a few weeks before Christmas, as I crouched in front of a fireplace filled with glowing coals arranging eggplant slices on a grill rack, it occurred to me that something had changed. I would never, in America, have been preparing to cook most of the dinner in the fireplace. Nor would it be a four-course meal whose centerpiece was *magrets*—duck breasts from birds raised not fifty miles away and force-fed for foie gras, which gave their flesh a silky richness best consumed (horror of horrors!) very rare. Six months ago I would have thrown them in the oven and cooked the hell out of them.

My wife, Amy, was in the kitchen putting together the salad, and my daughter sat on the sofa riveted by a French nature program about bats, a glass of bilious green mint syrup mixed with water—her *apéro*—in front of her. To my right, on the chestnut bench built into one side of the very wide hearth, sat two bottles of 1990 Château Chambert, the best recent year of one of the best local châteaux, which would be served *chambré*, at room temperature. Not only could I tell you the three superior vintages of the last decade or so (for *vin de Cahors*, the local AOC, 1990, 1995, and 2000 and the worst, but I could do a passing imitation of a wine waiter at table, that is, more or less gracefully uncork and pour, then taste and describe what was in the (appropriate) glass.

And so could my wife. Even Lily knew to look at the end of the cork and sniff it, then swirl just a bit of wine in the glass before sticking her nose into it and inhaling deeply. We would sometimes play a game at dinner, all of us coming up with adjectives to describe the wine's smell and taste, Lily being allowed a tiny sip. *Charpenté* (full-bodied, chewy) or *fruité* (lighter, more fruit and less tannin) we could agree on, but then came a subjective flood of berries and fruits that quickly deteriorated into the realm of spontaneous association—dirt, truffles, oak, leather, tar, *sous-bois* (leaf mold), grass, wet wool, goats (Lily), and just wine (Lily again).

Any single bit of behavior like this would—and, later, did—get us condemned as food and wine snobs at home, but here people looked at you like you were an idiot if you didn't know such basics. In this environment, in which the awareness of what you eat and drink is as normal as breathing, we had, in doing nothing more than observe our neighbors and shop the markets, asking all the obvious questions as we went, begun to internalize a whole body of knowledge that was, in turn, intimately related to a way of life we were beginning to live.

We were apparently in danger, as my visiting brother had said only half mockingly, of becoming "Frenchified." As I surveyed the room, the evidence of a different approach to life and its living was everywhere, from the mundane to the significant.

The dining room table in one corner of the single large, downstairs room was set for four, Jacques and Noëlle being expected for dinner later, after Lily had gone to bed. I could hear my wife working in the kitchen around the corner, poaching slightly underripe pears in butter, cleaning mâche greens, and crumbling Roquefort, all of which would go into the warm salad. The entrée, Piemontese *bagna caôda*, I had prepared earlier, sharp endive leaves filled with slivers of smoky roasted red pepper which would be drizzled at the last minute with a piping hot sauce of minced anchovies and garlic gently softened in butter and olive oil. The eggplant slices I had marinated earlier in olive oil, balsamic vinegar, parsley, and lemon juice. Grilled, they would accompany the *magrets*, most of their fat removed, what was left on the skin side scored and rubbed with a bit of *gros sel* and black pepper and then fired just rare over the hot coals. On the sideboard lay a perfectly ripe tranche of Tomme de Vigan, a hard, nutty goat, for the cheese course, and Amy had made a simple, rich, but very light lemon ricotta cake for dessert.

Lily's metamorphosis had been the most obvious in her now full-scale assault on fluency, but at the same time she was also absorbing those myriad bits of culture, behavior, ways of being, everything that is summed up in that single French word, *la civilisation*. While her sentences were ever-growing in complexity as she soaked up new words, more telling were the small, everyday things, like the fact that she knew what an *apéro* was and had taken to drinking the kid version before dinner.

From her first day of school in mid-September, she had slowly transformed herself, chameleon-like, into a little French child I sometimes did not recognize as my own. In the beginning, we drove her to school and back in Marminiac on the other side of Cazals about fifteen minutes away, one concession to the rigors of a school day which began at nine and didn't end until four-thirty. Another concession was allowing her to watch French cartoons in the morning before school, television being something she'd been denied at home. (As a result, today she knows the Rugrats as *Les Razmoquets*, but also *Les Aventures de Princesse Ceci*, a lushly animated period drama set in the seventeenth-century Austrian court in which the eponymous heroine finds herself, each week, at the center of political turmoil, social upheaval, and outright revolution, all based on actual historical events.) The school week was four days, and went nearly year-round, with long vacations four times a year and seven weeks off in the summer. It was very taxing, especially in another language when absolutely everything was new and thus the source of much trepidation, and not only for her.

The French educational experience at this level held many surprises for us. Lily was in the *maternelle grande section*, the second of three preparatory years starting about age four before kids move to *classe primaire*, or first grade. Having seen only slightly older kids staggering under the load of their gargantuan bookbags, we were stunned when we found out, in a letter from her teacher, that the only thing Lily needed to bring for her first day of school was a cloth napkin with her name marked on it. She could also bring a *doudou* (her blankie, used to *faire dodo*, to take a nap) and a stuffed animal if she wanted, but, some things being international, Pokémon cards and toys were strictly forbidden.

The first week, she put on a brave face every morning, then burst into tears just as we walked into the schoolyard, clinging to her mother or to me until one of her teachers would come to reassure her. There is nothing so terrible as turning your back on your own crying child, leaving her to the mercy of you know not what, and we wondered if we were doing the right thing. But there she was, at the end of the day, happy, if a little quieter than her usual boisterous self.

Driving home, I would see her in the rearview mirror, curled up in the corner of the backseat, Isabelle stretched out beside her, Lily caressing her head as much to soothe herself as the dog. I would ask her, in French, what she had done in school that day, and she would melt into a pathetic puddle of six year old before my eyes, almost in tears. "Dad, I had a hard day," she would say dramatically. "I don't want to speak French!" Then, as we approached Cazals and the moment when she could choose a single piece of candy at the tobacco shop, "Can we go to the *tabac*?" When she did deign to open up, it was a recitation of the disasters and misfortunes of the playground, who cut their hand and bumped their head, who wouldn't stop talking and had to go *au coin*, to sit in the corner.

At first we heard mostly about Daniel and Alexandre, the two English boys, then, as fall wore on, Noémi and Lucie and Océane and Luna started cropping up. Lily spoke often of Martine, too, the sturdy, comforting teacher's aide, who managed, with minimal English, to calm her fears and comfort her, and Liliane, her older, gentle voiced teacher. This was in marked contrast to Sophie, the younger, stricter teacher whom we heard bellowing one day through an open window, "HUGO! You don't listen to a word I say!" Martine reassured us that Lily was mixing more and more with the French girls and progressing in leaps and bounds.

Soon Lily started bringing home French words and phrases we had never learned: *gribouiller*—to scribble, *on va aller dire de toi!*—We're gonna go and tell on you! She began to sing songs, too, some of which we thought she hadn't quite mastered because we couldn't understand the words, until we found out they were in patois, which was what some of these kids were speaking at home.

While she seemed unable to describe what she was learning at school, she often spoke of lunch and recess. It was the first time I noticed that she was absorbing large numbers of French words for which she had no English equivalents, *un feutre* (felt-tip pen), *la cabane* (playhouse), and *la vinaigrette*, for example. She was also learning the vocabulary of French cuisine. "We had a hard-boiled egg and tomatoes for entrée," she told me one day, "*saucisses et pâtés* for the *plat*, and a *fromage* I can't remember what kind, and *gâteau* for dessert." Yes, my child was eating, during a leisurely lunch hour and for about a dollar fifty a day, a three- or four-course meal. This one-hour lunch, by the way, was followed by nearly an hour of play or rest time, so the children could digest before returning to schoolwork. American first grade, with the twenty-minute snatch-and-grab that passed for lunchtime, would be a rude shock to her.

The French do not seem to think, as we do, that young palates need to be coddled with the bland concoctions of the typical American school lunch, macaroni and cheese, hot dogs and hamburgers, french fries, and the like. During the course of the year, Lily ate sardines, anchovies, *brandade de morue* (salt cod and mashed potatoes), various charcuterie, roasted chicken and duck, stewed rabbit, turkey breast, pot-au-feu, *blanquette de veau*, pâté de foie gras, rillettes, stewed lentils, potatoes dauphinois, eggplant, stuffed zucchini, ratatouille, couscous, crudités, beet salad, grated carrots in vinaigrette, endive with tuna, and cauliflower with white sauce, to give only a brief list. (Hence the all-important napkin.)

How interesting—how natural, too—it was to see the seeds of gastronomy sown so sensibly in newly turned, receptive ground.

Although we hear about how quickly small children learn
other languages, when your own child begins to take on a different
persona under the influence of that language, it can give you pause.
Partly, it is the sheer rapidity of the process. At the Café de Paris,
on those first hot summer afternoons, almost the only words Lily
could say were *un jus de plum, s'il vous plaît*, addressing Sandrine
or Isabelle, the two young waitresses who were the first in what
became a crowd of locals who would all pitch in to teach her
French. (She was trying to say *jus de pomme*, apple juice.) She
knew the words of politesse, and she could count to thirty thanks
to eight-year-old Melvin Laval, but that was the extent of her
French. Four months later, at the end of October, at Noëlle's family gathering for Toussaint, she stood on a chair after dinner and
belted out a word-perfect if not particularly tuneful "Au Clair de
la Lune." Two months after that, she would toddle off, all alone
for the first time, down the road to spend an hour chatting with
Suzanne Laval, who speaks not a word of English.

When she began to teach us two or three versions of *plouffe
plouffe*—eenie-meenie-miney-moe, even the concept of which she
had never learned in English, that was a milestone. (One of the
naughtier versions went: *Est-ce que t'as vu une femme toute
NUE? Quelle couleur était son CUL? ROUGE! Est-ce que t'as de
rouge sur toi?* Have you seen a lady all naked? What color was her
ASS? RED! Are you wearing RED today?...)

Like every six-year-old, Lily had her bad moments in English,
but it was a bit perplexing to have a little French brat throwing a
fit and screaming, *Nangh! Je veux pas! Papa, tu es le caca d'une
chèvre!*, and refusing to get out of the car one afternoon. I was not
bothered that she spoke with the southwestern accent (hence the
twangy *nangh* in the place of the nasal standard French *non*), but
having her tell me I was goat poop, and with a mien and intona-
tion exactly reproducing a fit I'd seen one of her friends throw at
the feet of *her* parents a few days earlier, stopped me in my tracks.

She lost her first tooth in November, an event that happened entirely in French. When she got into the car after school that day, she held out an envelope containing the treasure and launched into an explanation: "*Papa, j'ai fait tomber une dent et la maîtresse dit que La Petite Souris, elle va venir et elle va laisser une pièce de monnaie et je veux aller au tabac et . . .*" She drew a breath and abruptly switched into English. "And can I buy some candy with the money the mouse leaves me?" Thus we learned that, here, it would not be the tooth fairy who put five francs under her pillow that night but La Petite Souris, the Little Tooth Mouse, just as the teacher had told her.

At Christmas, when we walked into Michelle and Alain Laval's house for a holiday meal, we saw a row of shoes lined up in front of the fireplace. Those are for Père Noel to fill, Michelle informed a puzzled Lily, which occasioned some quick footwork on our part to paper over the total absence of reindeer, sleigh, Christmas tree, and stockings. And at Easter, it was also Michelle who would ask her what *les cloches*, the flying bells, had left her for Easter treats. The Little Tooth Mouse, Father Christmas, flying bells—not exactly towering cultural icons, you would think, except that she was experiencing these childhood landmarks, some for the first time as a more conscious little girl rather than an oblivious toddler, outside her native language and culture.

Moments like these provoked in me faint stirrings of chauvinism, the whisper in the back of the head that says, "If you stayed another year, your daughter would be French. Now, is that a good thing? Is that what you want?" We had noticed the same tensions, more clearly manifest, in the few English couples we knew, especially those with older children. "She makes a right cock-up of it when I ask her to write a letter to her gran in English," one mother said of her teenage daughter, not at all happy that the girl was losing her ability to write in English.

And language could become a cruel weapon, too, when children spoke French fluently and their parents much less so. "You just called me a chicken, Mum. I *think* you were trying to tell me to shut it. Right, Mum?" This is what I heard a snotty teen say to her mother one day in the café as the two fought over neglected schoolwork, and the mother attempted, speaking bad French, to get her daughter to lower her voice. The daughter was embarrassed by her mother's poor language, embarrassed to be upbraided in public in front of her school pals, and struck back with the only weapon to hand, her tongue. After insulting her mother to her face, she then insulted her again to her friends in a rapid and colloquial French she knew her mother would not understand.

We realized, of course, that our turn would come, too, if we stayed on here, because Lily had already reached the point where she would have a facility, an effortlessness, that we would never attain. By the time we left, Lily could and did change languages like she changed her clothes, several times a day depending on the company and her mood.

Language is something my wife and I are both naturally drawn to (learning them and throwing the javelin seem to be my only God-given talents), and our French, so dilapidated on arrival, was more or less fluent if not elegant after six months in the country. Yet we would always be handicapped for the simple reason that it was not until the age of fourteen that we had both begun to learn it, too late for it to be hardwired in our brains. I can speak well and volubly for three or four hours, but a new and demanding situation, especially one requiring sustained attention—spending a morning with a winemaker learning her operation and then doing an in-depth interview for three hours in the afternoon—leaves me limp with fatigue. When the ringing phone summoned me from sleep at seven A.M., it took a conscious effort to summon the *"Oui? Bonjour!"* and then carry on anything resembling a coherent conversation. Translating the words of a tour guide or a wide-ranging dinner party conversation for our visiting parents, both of which require rapid changes of language every few minutes, these were tasks that, while we could pull them off, left us mute and with pounding heads after an hour.

Our questions about Lily's future had no answer, but no real urgency, either, for we were not going to live here forever. Many of our English friends were permanent residents, however, and I could see that these issues were difficult for them, threatening in a visceral way, especially as, with each passing year, they watched their children become inexorably and irretrievably more French, gradually losing their ability, despite their parents' efforts, to read and write easily in English and missing out on those key experiences that would define them clearly as English one day.

Language is, of course, insidious (as is the culture it communicates), permeating the most mundane of social transactions. If you do not function at a high enough level, you exist in a degraded universe, deaf to the life around you and mute in response to what you *do* understand. This was the fate of many of the expatriates and foreign retirees we met here, especially for some reason the English, their inability to communicate dooming them to a limited and sometimes very narrow view of France. A common complaint was that the French never invited them into their homes, even after a long association. "But what would we talk about? *How* would we talk? They don't speak French!" came one simple explanation, delivered in slight exasperation, by one of my neighbors.

When you climb inside another language, draw it around you like a second skin, this cloak of fluency allows you not only to blend in but to participate fully in all those prosaic, almost random experiences that make up a rich, full daily life. I'm talking about the chance encounter on your afternoon walk that turns into your neighbor showing you the hidden spring where you can harvest watercress even in February. This is the invitation to a pig slaughter delivered over the gas pump, the compliment you give the goat cheese lady at the market returned tenfold when she says, "Come and see the newborn kids in the spring!" Such experiences make you feel that you belong, that you are that much less a foreigner, a stranger, a visitor, or an observer, all of which can be soul killing if they are the sum total of your existence.

After the magic of the newness, the landscape, the sensual delights wore off, after, in fact, summer ended and all the vacationers and tourists went home, we were left in a world in which the normal signposts were missing. We built our life at first of small things, going to the bakery in Cazals on the way to take Lily to school, taking long afternoon walks around the village and its environs, picking up the newspapers at the *tabac*, stopping for a café au lait or a glass at the Café de Paris on the way home from school. As the only village for miles in which to buy everything from the bottled butane that fired our kitchen stove to meat, bread, newspapers, and lightbulbs, or to have a lively, inexpensive lunch at the Auberge de la Place, Cazals in particular became a place where we spent many hours in a week.

Over time, such sparse beginnings, if properly nurtured, can lead to greater things. In the small villages of the Bouriane, the foreigner who stays on after the first of September is an oddity, and you earn a few points just for having chosen to throw your lot in with the locals through the long, often dreary winter. At the Sunday market in Cazals, which had lost half its stalls with the changing season, Madame Stuker, according us this new status, began to favor us with her advice on choosing from her vast selection of fruits and vegetables. "You want those pears?" she said one November day, her nose wrinkling, "I don't think so." She leaned forward conspiratorially. "They're a bit . . ." Her lips turned up in distaste, and she waggled her hand, signaling secret knowledge of their age or provenance. "Some apples, perhaps?"

I knew we were making some social progress when, toward the end of October, the kind Madame Crochard at Bruno Bouygues' bakery (the best of the two in town) offered that Lily was making great progress in her French. I thanked her both for the compliment and for her patience, as Lily insisted we wait outside the door each morning while she bought the bread herself, a sometimes lengthy process since she usually tried to cadge (sometimes successfully) a *pain au chocolat* into the bargain. "Monsieur, you might want," she went on, "to reserve your *pain* for Sunday and Monday because, you know, the Parisians are back for the holidays." I must confess to feeling a warm glow at having my custom valued over that of a Parisian, though this might have been due to the pushy, boorish reputation the latter had out in the country. (Madame Crochard was not only kind, but quite lovely in an earthy, vigorous kind of way. I once heard an old codger compliment her, saying, "Why you, you're bursting with health from every hole today!")

The never-ending U.S. presidential election proved a source of great puzzlement, and great fun at our expense, to the French, and not only in its length. "Pardon me, but isn't he, unhhh, a bit simple, Monsieur Bush?" "What is it about you Americans? You're going to replace a sex fiend with either an idiot or a bore!" The election also gave people reason to venture a little more deeply into conversation with us.

Monsieur Alazard at the *tabac* began a little ritual exchange when I came each day to pick up the paper he held for me. When he saw me walk in, he would lean forward over the counter, my paper in hand, shaking his head gravely as if all the news were bad. He would peer over his bifocals at me, pause dramatically, and then say, "So sorry, monsieur. You STILL don't have a president." Then, with a superior sort of smirk playing about his lips, he would lean back, drawing himself up to his full, five-feet-six-inch height, and add encouragingly: "Perhaps tomorrow?"

Almost every person, every establishment, every place around us had a story, most of them quite interesting and always revealing of the way things worked out in the country. I found that, if you just showed up often enough, if you could listen and understand and give the minimal responses required (No! Really? He said that? What a shame! Some weather we're having, eh? Terrible news about . . .), you would eventually hear them all, some of them over and over again.

After a few months, I became convinced that if you sat for just a few days in the Café de Paris, which takes up a whole corner opposite the big open Place Hugues Salel at the center of Cazals, you would eventually see every man, woman, child who lived for fifty miles around. It was, I concluded, a rather peculiar institution and it had its own complicated story, too, which began with the supposed wealth of its owners, who were so frugal they only lit up the café's magnificent red neon sign twice a year, at Christmastime and on July 14, Bastille Day, France's Independence Day.

The café consists of a large open room with a bar at one end and worn red Formica-topped tables over wooden floorboards that have long ago lost their varnish. There are loving cups and trophies on high, dust-covered shelves, the ubiquitous cardboard Tarif des Consommations and Notice to Minors and Against Public Drunkenness on walls tinged yellow from decades of cigarette smoke. There are doorways at either end of the bar, one leading to the butcher shop next door, which Monsieur and Madame Bouyssou also own, the other to their dining room in the back, and from which blasts the television because Madame is rather deaf.

There are two glass doors in the back wall, too, leading to another room with tables and chairs. This room is usually empty but for racks hung with Madame's drying sheets, and the occasional youngster banging away at *le flipper*, the pinball machine. On the second Sunday morning of every month, however, the curious will see a serious-looking group gathered around tables set in a square, smoking furiously and consuming endless espressos while engaged in sometimes heated debate. This is the *café philosophe*, a group of a dozen or so neighborhood intellectuals who gather to discuss, with great erudition and not a little pomp, such eternal questions as What is Beauty?, the topic being announced the week before in the local newspaper so the members can do their homework. It is also a group I never had the courage to join, the few snippets I overheard ferrying Lily to the bathroom in the rear transporting me to the year I had spent at the Sorbonne, usually at the back of the lecture hall, desperately trying to understand what the hell the professor was running on about.

Madame and Monsieur Bouyssou, who is only a little deaf, are most likely in their early eighties although no one knows for sure. They are both short and stocky, both wear the uniform of the shopkeeper, she a white cotton coat like a doctor, his of pearl gray, collared shirt and tie underneath. By trade she is a butcher and he a charcutier. They own the café, the butcher shop, and the shuttered, twelve-room hotel above. Underneath there is a state-of-the-art charcuterie and butcher's workspace and apparently the most spacious, impressive cave in the whole village. They don't smile a lot, and, on first impression, might appear to a stranger out of sorts.

I found them difficult to talk to and felt bad about it because, between her hearing disability, his dentures, and the thick accents and bits of patois peppering their sentences, they were quite simply very hard for me to understand. This was unfortunate, for they were certainly not shy, and happy to chat once they knew you a bit. Madame fell instantly in love with Lily, was always, in fact, inviting her behind the bar and into their dining room, exchanging a cookie for a *bisou*. The two would return, hand in hand, Madame marveling to me over Lily's increasing fluency and gabbing on about her own children. We could never get much beyond that, however, because she couldn't hear my replies. Monsieur, portly, watery blue eyes over a generous nose and a mouth shiny with goldwork, was another story, his usual conversation revolving around the single topic of his new house and the cabal of contractors who had conspired to screw him over. The couple's occasional gruffness was, however, completely mitigated by Sandrine and Isabelle, the two lively bargirls who ran interference for their employers, smoothing over their rougher edges, flattering the old men, welcoming the young, ready with a smile and a kiss of greeting for one and all. They were truly two bright lights in our small universe.

To the outsider, the social order of the Café de Paris might at first seem impenetrable. It certainly did to us. What seemed like an unending stream of people would walk in off the street, get a *bisou* on each cheek from Isabelle, Sandrine, and anyone else within kissing distance, and hearty handshakes all around from the men, and be immediately sucked, Hoover-like, into what always sounded like some fascinating conversation. In the summer, our welcome began with a garden-variety *bonjour* from Monsieur or

Madame and ended with a once-over from the others, who then turned their backs to resume their conversation. We were taken quite naturally for tourists.

When I began to turn up several afternoons a week once Lily was in school, however, and when I mustered the courage to belly up to the bar to take my café au lait or *demie* of beer shoulder to shoulder with the locals, it was the start of a slow progression toward, not exactly intimacy, but familiarity, at least. We had long ago introduced ourselves to Isabelle and Sandrine, and soon we found ourselves on a first-name basis with them, kissing cheeks over the bar. The formal *vous* gave way to *tu*, and the younger group, all in their twenties, began to shake my hand, some to greet me by name. Little conversations about the weather (wet, raw, weeks of rain that winter), grew more substantial as I paid more attention to local news and events and asked questions about them. I always learned some surprising things.

Early in December, for example, I walked in one afternoon to find Philippe Moreau, whose magnificent waxed mustache and bushy red muttonchops give him a somewhat fierce air, telling the tale, his voice rising as the outrage unfolded, of a dastardly criminal operating in the area, the foie gras thief. Philippe is an *éleveur* and *gaveur* who runs Les Cabanes, a duck farm and foie gras operation in nearby Montcléra, selling his goods at the Cazals market every Sunday.

In four neighboring villages in the last year, someone had burgled farm storehouses like his, making off with hundreds of pounds of foie gras, thousands of duck breasts, canned gizzards, pâtés, the lot. This last outrage had taken place the weekend before in Frayssinet-le-Gélat, ten kilometers away. Often the thieves strike just before the *éleveur*'s market days, when he has just butchered and his larder is full. This represents months of labor and tens of thousands of dollars, and these are not people who can afford to insure against every risk. Often, too, the thieves steal products that need immediate processing, which requires equipment. This, along with the timing, suggests that the thief is, in fact, one of their own, which has made all the *éleveurs* uncomfortable and suspicious. Some have taken to sleeping with loaded rifles at their bedside and swear to shoot the miscreant down for stealing a man's lifeblood.

"This farmer's *conserverie* [where he butchers, processes, and cans] is not only in the center of the village, but just a few meters from his house," Philippe spluttered, growing ever redder in the face as he related the tale, "with dogs all around, even. But nobody heard anything! It was disgusting, the thieves left sausage flesh strewn all over the floor and made off with one hundred twenty breasts in salt, two hundred fresh gizzards, and a load of foie gras. I'd shoot the bastards! Put me in jail, go ahead." He turned to me, saying, "It's illegal to shoot anyone in France, even in your own home, unlike in many of your states. I don't care, I'd still shoot them!"

The talk centered not only on the audacity of the thief, but the towering incompetence of the local gendarmes. "And the cops can't seem to find their own asses, never mind this guy," a man on a neighboring stool piped up. "They can't even catch the guy who breaks into Casino [Cazals' superette] every Christmas holiday. You know about that?" He looked at me. "For the last few years, every Christmas someone has broken in, helped themselves to foie gras and Champagne, then left. They eat and drink right there and leave the trash! And the police station is right behind the store."

It was also in December that I learned over a beer one evening that the café, the butcher shop, the hotel, the whole place was for sale. Madame hinted darkly that they were going to close in January, Monsieur avowed as how he was ready to retire, and it seemed a done deal. Then it emerged that this was a yearly ritual with the two, at the holiday season to begin to make low, grumbling noises about shutting down, selling up, retiring. This was why Monsieur was building his luxurious retirement house down the street, why the café hours, always erratic, had become ever more unpredictable of late. Perhaps this was also why, I thought, the two certainly seemed to make no effort to coddle their customers. They served no food, not even a basket or two of morning croissants or tartines, certainly not a sandwich at lunch, even though they could have done a roaring trade, especially on the weekends and in the summer. But no, they closed early in the afternoon to have their lunch, sending the English, who were often settling in for a good, long booze-up, down the road to Pascal's Bar in Salviac.

This was, of course, only half the story, for Madame hated the new house and was apparently refusing to move when (and if) it was ever completed. When Monsieur was off on one of his tears, (and there were many, about a terrace that drained rainwater toward and not away from the house, about two-inch gaps under the doors, about his good topsoil stolen during construction and replaced with fill that wouldn't even grow grass, about pipes bursting when taps were opened for the first time) Madame looked on, conspicuously silent with her sympathy. The price, too, variously reported at from three to five hundred thousand dollars, was far too high for the market. "Outrageous! The rooms upstairs haven't been redone since the forties, no bathrooms, no heat! That'll cost millions of francs to redo. What is he thinking!" This led the locals to grumble that it could only be some fool foreigner who would buy it, or that he wasn't really serious.

And then Madame closed the butcher shop.

A few weeks after Christmas, I had noticed the blinds drawn each time I went in to Cazals to do my shopping. (We patronized the other butcher in town, or we would have known earlier). I was at the greengrocer's, and asked Lydia, the proprietress, if the news was true, that Madame had closed up shop. "*Oui, oui.* She said it was time, she was just tired." But, I persisted, there's a meat truck pulled up in front of her store right now, and I thought I saw someone coming out of the café with a package the other day? "Oh, well," Lydia continued, "she didn't give up her license. She's still selling meat to her old customers. The old lady who just left, with the red cabbage? She comes down from Villefranche-du-Périgord! To get her meat from Madame and her vegetables here!

But she's not selling to all of her old customers," she finished, sounding a bit put out. "Claude [Lydia's husband] does like his red meat, and you know you can trust them."

The meat truck had pulled away by the time I entered the café and took a seat a few minutes later, but there was Madame, looking unlike I had ever seen her before. Gone was the apron, in its place a fancy dress and scarf above and shiny shoes below, the whole crowned by a new coif of stiff brown waves. Most surprisingly, she sat at one of her own tables, a cup of coffee in front of her and chatting nine to the dozen with the old lady Lydia had mentioned, who had the most amazing brilliant purple hair and appeared as if she, too, had gone to some care with her appearance. This madame sat across from a gent of the same vintage with one knee-braced leg stretched out in front of him and crutches at his side.

"There was something grating in there," the gent was saying, tapping his kneecap, "something terrible it was, for weeks. Then, last week, some kind of pimple raised itself, all red and ugly, then it burst!" The two women hung on his words, nodding sagely, their eyes following his hands as he mimed a mound rising and exploding out. "So I went to the doctor, who sent me to the specialist, who sent me straight to the hospital. And here I am!"

The talk then turned, in a rather bizarre segue, to meat, specifically to the recent discovery of *vache folle*, mad cow disease, in the French beef herds. While England had already slaughtered tens of thousands of beasts (and suffered a handful of deaths in those infected with the brain-wasting human equivalent from eating infected meat), the French had gone to bed each night assuring themselves that this was a (fittingly) British problem. The French government had a longstanding embargo of all English beef, and the famously rigorous French veterinary inspection services and food inspectors stepped up their surveillance to make sure the butcher shop was safe for every madame and monsieur in pursuit of a fine roast.

In the fall, all that changed overnight with a strict decree (fought tooth and nail by the beef producers and processors) vastly increasing the numbers of animals bound for slaughter to be tested. Instantly, the number of cases detected doubled, then tripled. Meat from suspect cattle was even found on some shelves at *les grandes surfaces* (shopping center supermarkets), seeming to confirm people's suspicions that these huge chain stores could not be trusted. School cafeterias stopped serving beef overnight, and the sale of organ meats and certain bone-in cuts was outlawed (no more *ris-de-veau*!). At La Récréation, orders for the perenially popular (and beautifully bloody) *pavé de boeuf* declined so precipitously that Jacques dropped it from the menu while the wholesale price of fish skyrocketed, nearly precipitating a ten-franc rise in the price of the prix fixe menu.

"What can you choose at the butcher's anymore?" Pimple-knee's wife was saying. "I had to take a *gigot d'agneau* the other day—from New Zealand! New Zealand!"

Madame Bouyssou, herself a butcher after all, clucked her tongue and nodded her head in sympathy. "*Eh, oui!*" she said. "*Eh, oui!*"

"You don't know what to buy! Or where it really comes from! And in *les grandes surfaces*, well, the meat there, they wash it to make it more red, you know, some special chemical. You never know," she continued, wagging her finger. "And what about those two butchers they just arrested for fraud in, where was it, Souillac? Selling rotten meat! You can't even buy local beasts, either. Or chickens. The chickens hanging in the market even, they *wash* them, in running water. The skin is wet! You can't roast that!"

"It's terrible, just terrible." Her husband took up the rant, and it was his wife's turn to nod and shake and wag and cluck. "We went to visit friends for a week. What did we eat? Once pork, twice chicken, and the rest fish, fish, fish. And we weren't even by the sea! Imagine! What will we eat tomorrow? Eggs and cheese? Rabbit and chicken and duck from our neighbors only?" At this butcher's nightmare, Madame Bouyssou drew back in horror.

"Yesterday," the man's wife interjected, "I went to the market. What did I buy? A cabbage, leeks, a cauliflower. For the leeks, ten francs. Ten francs! And so little white and so much green! And cod! Cod! Cod! Cod! How much did they want? Ninety francs the kilo! No, ninety-eight francs. Ninety-eight! Ninety-eight! *Oh là là!* So expensive. So I bought the mussels instead. Me, mussels! And there was a line at the *boucherie chevaline*!"

This last news, that there was a run on horsemeat (yes, they still eat horsemeat in France, where it is sold by special butcher shops displaying a horse's head on their sign), seemed to be too much for Madame Bouyssou, who didn't know whether to nod her head or shake it. *"Eh, oui,"* she finally said sadly. *"Eh, oui! Eh, oui! Eh, oui!"* The conversation veered off into other things, and finally the couple got up to leave. "Oh!" Madame Bouyssou said. "Don't forget your meat!" She disappeared through the connecting door into the butcher shop and returned with a hefty parcel wrapped in butcher paper. "Thank you," the other woman said. "Thank you very much!" And she really meant it. The visit aside, she and her husband had driven almost two hours round trip just to get their meat from an old friend, someone they trusted.

I saw more of Madame Bouyssou out from behind the counter in the following weeks, but things, I heard from Isabelle and Sandrine, were not going well. Madame, bereft of her normal routines, seemed adrift and began to suffer from headaches. Docteur Martin, who could often be found having an espresso at the bar when his office wasn't busy, did not seem able to help. Madame went to the local *guérisseur*, a faith healer, but his efforts also had failed.

Then one day she was back in her lab coat, shuffling from butcher shop to café, a smile on her face once again. The shades were up in the butcher shop, the café seemed once again to have been spared as well, and all was right with the world. "I never meant to close," she told me later. "It was just a break, a rest!" The real answer, I suspect, was simply that, deprived of her main avenue of social interaction, she was bored plain and simple, and so used to working that not to work was anathema.

Americans sometimes think that, because cafés serve alcohol,

and because people drink there, they must be the French equivalent of bars. The primary function of the café, however, is social, a
meeting place, neutral territory to retreat to, even for five minutes
before lunch. I never, in thirteen months, saw anyone drunk at the
Café de Paris. You went there before work, or before lunch, or
before dinner to exchange the news of the day with friends and to
have a beverage. The most popular item they served, Sandrine told
me, was espresso at eighty cents the demitasse. When people did
have a drink, it was most often a single glass of wine or muscat or
pastis. Sure, I often saw a table of old men sitting over their glasses
of red wine at eight in the morning, but they would not be totter-
ing out, drunk, three hours and five glasses later.

For us, the café played an important role, a social half hour at
the end of many a solitary day when we might see some people
more nearly our own age and with whom we could talk. For, as we
realized when the cold rains of fall rain set in, there simply weren't
very many people out and about in the countryside where we
lived, and almost none between the ages of twenty and fifty. The
average age of the friends we had made, we calculated one day, was
sixty. Our social anchors, Jacques and Noëlle, not only had their
own lives and obligations, but most times were working. In the
whole of Les Arques, I could count only ten people of our generation, and even fewer of Lily's.

The substance of my days was dictated partly by the seasons
and partly by my own interests. Foie gras was traditionally a fall
activity, winter the time of truffles, spring for planting the fields,
summer for the first harvests. I made my own schedule, choosing
to spend time with this farmer or that winemaker when his work
was interesting. Because part of my original intent was to follow
the restaurant, too, through the seasons, I would haunt the kitchen
for at least two weeks during every three-month period. Jacques
was good enough to alert me when the unusual and interesting
came up, grilling a whole boar for the Hunters' Supper, for example. (They ate the heart and liver as well as the meat.) I pitched in
washing dishes and doing the simplest prep every weekend in

October and November, washing mushrooms, peeling squash, and chopping vegetables for soup; cleaning scallops, shrimp, and red mullet; turning sheaves of parsley, basil, chervil, and mint into garnishes.

Between these bouts of activity, you would have found me in front of my Macintosh laptop at a small desk in one corner of the dining room. My wife was working on her own book, her laptop on the dining room table ten feet away, and there were whole weeks that passed with the two of us sitting there, not saying a word except "Lunch?" or "Walk the dog?" or "Go get Lily?"

Nearly every school day, Amy would get Lily ready and drop her off, then disappear with the dog, sometimes for hours. She and Isabelle became such a ubiquitous sight in the countryside that, when I was walking the dog out in some distant hamlet, I would often come across people who knew her. "Oh! The woman with the dog." That's what they called her at first, the woman with the dog.

In late September, however, she gained another sobriquet, *l'américaine qui a planté la voiture dans le pré de Bousquet*. That particular afternoon, she had left our first car, a four-door Peugeot 405, parked just around the corner from Bernard Bousquet's house up in the village. We had a rendezvous with some friends and were standing around on the steps of their house, chatting. "Mom, the car . . ." Lily, in her normal motormouth mode, interrupted. "Don't interrupt! Please wait until we're finished talking, Lil." Amy said, turning back to her conversation. "Mom, the CAR!" Lily repeated.

Hearing the urgency in Lily's voice, Amy stepped out into the road to see the back end of the car vanish through an opening in a stone wall bordering an open field. She ran down the street, turning into the field just in time to see the Peugeot embrace a tree with a terrific crash. We went up to the restaurant to call a wrecker, precipitating an exodus from the kitchen, everyone eager to see the damage. The unusual activity drew other villagers, and the grapevine filled in those who weren't around. Within days, Amy had become "the American who totaled the car in Bousquet's field," and we found ourselves with a new identity which I'm sure remains attached to us this day. (And a new car, this one a Lilliputian Nissan Micra that, I used to joke, was named that because the next model up was the Micro.) Even months later, sometimes when I was introduced to a villager I hadn't yet met, it wasn't as "the writer" or "the friend of Jacques and Noëlle." It was, "You know, it was his wife who totaled the car in Bousquet's field." "Oh, yes," the other person would say. "I saw that car. Terrible bad luck."

Such momentary excitements aside, as fall melted seamlessly into winter (it rained continually) we both realized that walking had become, and probably would remain, our primary recreation. There just wasn't that much else to do. The café was closed at seven-thirty or eight each night, the nearest cinemas were all forty minutes away, there was no health club or a bowling alley, pool hall, or dance club. Most of the restaurants in neighboring villages weren't even open at night out of the summer season, and spontaneous invitations from our few friends were few and far between. Going out in the evening was also difficult because finding an English baby-sitter for Lily was hard, and a local French lass just too much of a stretch for her still.

We were finally beginning to realize what so many people had told us, that the Lot was a lonely place, thinly inhabited, beautiful but desolate, especially when the cold and drear set in. While my work took me off regularly, Amy began to feel more acutely the lack of social interaction with anyone else but me, Lily, and the dog.

It was about this time that we began to have "the talk." "What's your real estate reading today?" one of us would ask the other, usually in bed before sleep, another day passed in our usual pursuits. As we chewed over the various benefits and challenges in the tantalizing possibility of pulling up stakes all together, or perhaps just buying a second home, our sentiments veered wildly depending on mood, weather, and the small successes or failures of the day.

Though we would surely make more friends if we stayed, the Bouriane was not going to transform itself into a more lively place just for us. Lily might possibly have only a bare handful of friends in the neighborhood. In the surrounding villages, and all over the

Lot for that matter, the demographic charts all show the same gaping hole where the twenty- to forty-year-olds should be. And where there are so few of these, there are even fewer children. Could we make a life surrounded only by the aging widows and retired farm couples who were our neighbors? Would our families come to visit?

But then, the next days would hold a surprise. The sun would come out, or Amy would escape to Toulouse with a friend for a day. Michelle and Alain Laval would call to invite us to go skiing with their two kids in the Pyrénées over a holiday. Jacques and Noëlle would have us up to dinner, or they would whisk us off to her parents' house outside Toulouse to be fussed over and fed by her mother, Lucienne. The questions would be put off, the urgency fade, and life continue on.

EIGHT

IN SEARCH OF
THE BLACK PEARL

It's over; it's screwed, the truffle . . .

—RAYMOND MÉNAUGES, TRUFFLER,
GINDOU, THE LOT

*In 1906 my great-grandfather brought in one ton of
truffles from his trees. . . . We will never again see
those kinds of numbers, but that doesn't mean we
can't improve the situation."*

—PIERRE SOURZAT, DIRECTOR,
TRUFFLE RESEARCH STATION, LE MONTAT, THE LOT

*The researchers had good intentions, but they went
in the wrong direction, which now, thirty years and a
generation later, has led us nowhere. I have a lot of
fears for my son in the future and a grandson who
[will not] be able to continue in the business at all . . .*

—JACQUES PÉBEYRE, TRUFFLE MERCHANT,
CAHORS, THE LOT

Merde *but that's good!*

—JACQUES RATIER, CHEF, EATING TRUFFLES

THE TRUFFLER

"*Putain* but that hurts!" Raymond Ménauges said, drawing in his breath sharply and pausing to rub his aching hip. We were a half a kilometer up the gravel road from his house, in Gindou, on the way to one of his two *truffières*, or truffle grounds. We had been slowly climbing the spine of a low ridge, stopping occasionally as Monsieur Ménauges's words outran his vigor to such a point that he lost his breath. Though he is still active at seventy-eight, a lifetime of hard labor as a farmer, vintner, and *trufficulteur* (cultivater of truffles) on his family's several hundred acres of stony ground has taken its toll on his body.

On our left, to the west, the land undulated away in gray-green pasture and walnut orchards, rows of crooked, naked grapevines, with a single newly plowed field of clayey, terra cotta–colored earth gleaming wetly on the next rise. "Les Arques is just over the ridge," he said, pointing in front of us. Winter is the season of *la taille*, pruning, however, and in the corner of every field, it seems, the farmers ignite their great piles of grape cuttings and walnut branches and corn husk and brush, filling the valleys with clouds of noxious white smoke that obscure the views. Where I expected to see a cluster of stone houses with the lichened roof of the Romanesque church rising up above them, there was nothing but haze. "And Cazals." He swept his arm to the right, where I could just make out the newer, outlying houses of that much larger town. "It's a nice spot, our little corner."

Behind us, sweet-scented woodsmoke curled from the chimney of his comfortable stone house, chickens scratched in the barnyard, and rabbits destined for the stewpot scurried about in their hutches. The walls of the sturdy outbuildings were well pointed, every roof tile in place, and, unusually for the area, there was nary an old tire or pile of rotting clay roof tile or hunk of rusting farm machinery in sight. On a mild January day, with the sun shining, Les Granges, the name of this property, seemed neither little nor merely nice, but then the people who live from the land have a way of understating things, almost as a defense against tempting fate or nature.

Monsieur Ménauges is the brother of Suzanne Laval, our neighbor, and it was in her house a few days previous that we had sat down to dinner to discover a first course of generous rounds of her own foie gras, each crowned by a thick slice of pungent, coal-black truffle. They came from her brother's land, up in Gindou, she told me, and she'd served them because she knew I was curious about truffles. Did I want to go visit him sometime? Only he was a real *bavard*, a chatterbox, she warned good-humoredly, and would probably talk my ear off.

As we had set out on our walk to the truffle grounds a few days later, almost the first words (of a subsequent flood) out of his mouth were, "*C'est fini. C'est foutu, la truffe!*" It's over, it's screwed, the truffle. "A terrible year, terrible. We've seen almost no truffles. No rain in July and August, which the truffle just can't tolerate. That's the problem now, isn't it? We have found some. I'll show you, they're pathetic, all dried out. In 1935, I remember going to hunt truffles with my grandfather. If he didn't get ten kilos a *week*, he wasn't happy."

"Baaf!" he continued, disgusted. "I don't believe in the truffle anymore. I was a passionate believer for fifty years, more even. Now . . ." He stopped and looked off into the distance, then shook his head once and abruptly started forward again.

We had reached the first *truffière*, about one hundred leafless, lichen-encrusted Michelin oak trees planted twenty feet apart in regular rows on two acres of ground. Underfoot there was gravel grown over in places with patches of weeds.

His wife, Christianne, had gone ahead of us with the dog, Dolly, a small, curly-haired sheepdog who has been hunting truffles for them for many years. Dolly looked from her mistress to the bare ground in front of her, put her nose to the dirt snorting and sniffing, then moved on a few feet before returning to Christianne's side. "Search, Dolly! Search! Find the truffle. You know it's there. Search!" Her soft commands floated over the breeze as they moved on, empty-handed, to the next likely spot.

As we strolled up one row Raymond's steps quickened. "Look at that *brûlé*!" he said. "Is it not beautiful?" A *brûlé* is a burn, a circular or oval-shaped patch of ground under the tree where the nascent truffles, even though minuscule, somehow kill all the vegetation. It does look, indeed, as if someone has swept away leaves, brush, dirt, even the soil over the ring of exposed limestone pebbles, a phenomenon produced, rather, by the fact that, once the truffle kills all the weeds, the wind keeps the empty patches clear of debris. "But . . ." He waggled a finger under my nose, his eyes on the barren patch of ground. "When a *brûlé* like that produces no truffles—which it hasn't!—we're finished. Not enough rain," he lamented.

As we proceeded up and down the rows, he pointed now here, now there to a particular patch of ground, a recently dug hole. "There I found a huge one, almost two hundred fifty grams!" Or, "A few years back, we must have pulled a kilo from under this burn over here." I was also noticing gaps in the rows of trees, where one, sometimes several, had been cut down. "Sourzat, the *scientifique* from Le Montat, he told me to do that. That when it produces acorns which sprout into seedlings, rip it out, it will no longer produce truffles." He shrugged. "So I did. And when the trees which were good producers slow down or stop all of a sudden, cut off the lower branches, he said . . . which I also did," he said, pointing at all the scars on the trunks around us. "But nothing has worked. It's so discouraging. And after all the work I've done. And the irrigation, too. Sourzat says not to water in the morning or during the day or too early in the evening. I'm going to come out here at midnight or four in the morning? It takes hours to water one hundred trees. I'm too old to spend all night in the fields in July and August."

It struck me suddenly that those impassioned by the truffle (with whom I'd been spending much time lately) share the worst tendencies of both fishermen and golfers. Like the former, one is always hearing about the one that got away from the *trufficulteurs*:

"Last year, I took a real elephant, a two-pound truffle from this one spot over here!" As Jacques Pébeyre, a truffle merchant in the region, had told me, "I don't care how many kilos he says he pulled out of the ground if the truffles never make it to my sorting table. If even the least part of those stories was true, I'd be seeing more truffles for sale each year and not less and less."

And every golfer has that one honestly good or even excellent round in a summer of his usual terrible play that keeps him coming back, buying equipment, and continuing to dream. For the *trufficulteur*, the equivalent is the single outstanding season when the truffles are ripe, large and round, and in sufficient quantity that he makes what seems a large sum of money. Enough anyway to expunge the bitter memories of the previous five years of wasted effort when the trees didn't even yield enough to take to market. The average yield, by the way, in this part of France over the last years has been about four to six kilos per hectare (1 hectare is about 2.5 acres) per year, or roughly four thousand to six thousand dollars an acre depending on the market price.

As we continued our walk, the highly anecdotal, sometimes contradictory observations Ménauges had accumulated from his years continued to flow. "Let the scrub juniper grow. Work the soil lightly and at a shallow depth in the early years; don't touch it once the trees begin to produce. In a good year, where there is a *bon brûlé*, even the snow gets burned, exposing the bare ground. When there are lots of acorns, there are no truffles. If you don't have snowy winters, not only will there be no truffles, but no cèpes, either, in the spring. You want a good freeze to help them ripen, but not too much to kill them."

In two hours, we found no truffles, which was perhaps to be expected since he'd been out with the dog a few days earlier. On our way down the hill back to his house, he continued his lament. "Everything about the truffle is a mystery," he said. "Now, it's like I've believed in Santa Claus for fifty years. I no longer believe." He perked up when we passed his vines, neat rows of auxerrois (malbec), merlot, and tannat grapes planted in the percentages corresponding to the usual *vin de Cahors*, seventy, twenty, and ten percent respectively. "We made a wine of 11.2 degrees of alcohol this year. Not bad for *un vin de Gindou*!" This last he said with a small smile, his little joke, as Gindou, a hamlet without even a bakery, is not noted for anything at all, much less its half-dozen *vignes de famille*, family vineyards whose wine is destined to be consumed around the kitchen table (or sold on the black to neighbors), most often wetting the inside of a bottle only briefly as it passes from barrel to glass.

We also passed another stretch of waste ground crisscrossed by rows of white plastic cylinders sheltering recently planted truffle oaks. Hope springs eternal, I thought, saying as much to him with a question in my voice. "You have to try." he said almost plaintively. "You can't just let things disappear. I was a young man with a bit of land. I sweated, I labored, always thinking, 'In the winter there will be truffles.' I was wrong. But you have to try!" Given that these young trees would not begin to produce anything for perhaps another eight, even ten years, by which time Monsieur Ménauges may well have passed on, I wondered for what grandson or granddaughter he was leaving them behind. For surely as night follows day, those trees would produce a few black pearls, tempting another, younger soul into a whole lifetime of truffle dreams.

THE SCIENTIST

A few weeks earlier I had found myself on another exposed hillside planted with scraggly, evenly spaced trees, the ground underfoot consisting mostly of drab pebbles of the local limestone in which even the weeds seemed to be struggling. It was a crisp, blustery day in the middle of December, the dormant oaks, which grow only fifteen or twenty feet tall, moving their leafless limbs pathetically against the backdrop of a cloudy sky in dramatic flux, though for better or worse one could not yet say. An odd time and an even odder place for a harvest, it seemed to me, for it was hard to imagine anything growing here, much less what has been called the black pearl, the black gold of Périgord. The black truffle, more precisely the melanosporum or truffle of Périgord, is not, however, normal in any way at all.

Melanosporum is a winter mushroom that grows underground (hence it is a tuber) in symbiosis with the roots of certain kinds of trees. Little truffles are born in May, grow most rapidly in size during the month of August, and begin to ripen in late November after a good frost, with the last appearing in the month of March. They grow best in chalky well-drained soil of moderate to meager fertility and a pH of about eight, need rain in the summer but not too much in the fall, require dappled sunlight on the soil over them, and are extremely susceptible to extremes of either heat, cold, or moisture. And all of this may or may not be precisely true.

"*Trufficulture* is not the cultivation of mushrooms." Pierre Sourzat, the truffle researcher who had invited me out into this cold morning, had told me earlier in his office. "It is the cultivation of mushrooms on the roots of certain trees. These are the green oak, the pubescent oak, the Michelin oak, and the hazelnut tree." (In French, respectively, *le chêne vert, le chêne pubescent, le chêne pédonculé,* and *le noisetier.*) The melanosporum mycelium, the moldlike medium that holds the truffle spores, forms a coating around the microscopic rootlets of the trees, a peculiar biological unit, part mushroom, part tree root, called a mycorrhize. Scientists know that there is an exchange, most likely that the tree processes through photosynthesis substances that the truffle cannot make while the truffle makes available minerals like potassium that the tree cannot easily get from the soil. When the soil heats up in the spring, by some process not yet understood the mycelium migrates from the rootlets out into the surrounding earth in a fine layer, and from this layer of mycelium little truffles (*truffettes*) are born.

That, much compressed, is the sum total of what we know of this strange, valuable winter fruit.

This paucity of knowledge is not what one would expect, considering that the black truffle is at the apex of the gastronomic

Richter scale, an explosion of such divine voluptuousness and poetry in the mouth that it has inspired legions of French writers from Brillat-Savarin to Colette to sometimes fulsome praise. More important to generations of Lotois past, it used to be one of the single bright spots in an otherwise impoverished agricultural landscape. Always one of the most expensive foodstuffs pound for pound, today the truffle has become an even more precious and rare commodity, selling in the 2000–2001 winter market at more than three hundred dollars a pound on the *wholesale* level.

Sadly, all of those oh-so-tempting classic (and usually very simple) truffle recipes—truffles under the ashes, sautéed truffles, truffles simmered in Champagne, shirred eggs with truffles—have become mere artifacts of a bygone era as the cost of eating truffles as a dish (thirty to fifty dollars per person) has soared out of reach of all but the most well-heeled of consumers. As a result, there is no retail level, or a negligible one perhaps in a handful of shops selling luxury comestibles where the blackly gleaming truffle, sorted, brushed, and washed, will empty the wallet at upward of seven hundred dollars a pound. Most are thus consumed either in expensive restaurants or as truffle butter or oil, or in a sauce or charcuterie that makes the most of their distinctive perfume and flavor.

Up on the windswept plateau, Pierre Sourzat wasn't worrying about any of that. He was out with his dogs to find some truffles. As if cultivating truffles wasn't difficult enough, you then have to find them, which means mastering the art of training a truffle hound (or, more rarely, a young pig), itself no easy thing. Pierre runs the truffle research station in Le Montat not far from Lalbenque, the truffle capital of southwest France, and he is the practical expert on truffles and trufficulture and harvesting truffles in the Lot.

Although from his reputation I had expected a stuffy professor of a certain age fussing in his laboratory, Pierre in person turned out to be much more down-to-earth. It is also very hard to take too seriously someone who has a minute bit of tissue stuck to his upper lip where he has cut himself shaving and who seems to have as many happy Labrador retrievers as employees loose in the office.

Pierre, in his late forties, is rather taller than your average Frenchman, dashing in a lean, square-jawed, careless kind of way. His blue eyes glitter under a full head of brown hair shot through with silver, and he has a pleasant giggle that emerges when something pleases him. He smiles frequently, especially when holding forth about truffles, betraying in conversation a deeply curious but easygoing nature that must stand him in good stead given the challenges and setbacks involved in trying to cultivate something about which so little is known. He wears expensive clothes much abused by the necessity to tramp around in the mud in all weather, the down-at-the-heels country gentleman effect. This carries over to his transportation, a white Land Rover, which in the land of four-dollar-a-gallon gasoline makes a certain statement, one much diminished as soon as I opened the door to venture out into the truffle oaks with him and was engulfed by a wave of intense odor composed of equal parts wet dog, mud, and a third, more pungent, almost chemical smell I could not identify and about which I did not dare ask.

The well-worn interior was stuffed chockablock with high rubber boots and boxes of equipment, with the camera and camera gear of the professional photographer (his other pursuit), with all the detritus of someone who is "passionate, no, crazy about truffles!" Why? His family has been growing and harvesting truffles for four generations at least, he himself owns a few score acres of working trees, and, as he says, "In 1906 my great-grandfather brought in one ton of truffles from his trees. One ton!" The entire French truffle harvest (melanosporum, or black truffles) in 1999 was less than thirty tons . . ." We will never again see those kinds of numbers, but that doesn't mean we can't improve the situation."

As soon as we had climbed into the car, his two yellow Labs, Darius and Octavius, started whining and yipping excitedly from their cage in the back. "Don't cry, boys!" Pierre called back to them. "We're going to put you to work right away." We left the fields and orchards and barns behind (Le Montat is an agricultural college for future farmers) and bumped across a road or two and through a field before coming to rest on the edge of a well-established *truffière*, one of Pierre's experimental plots.

As Monsieur Ménauges had shown me, you don't look for truffles randomly in a truffle ground, but in a *brûlé*. These round, oblong, or oval burns occur under the trees where the truffle, by exuding a kind of natural herbicide, gradually kills all the vegetation except in places for the moss, the only green thing, other than the lichen on the tree bark, to be seen for miles in this kind of country at this time of year.

Brûlés are quite distinctive (even to my novice eye), and begin as small as a few feet in diameter but may spread in a wide circle dozens of feet across. The burn expands or contracts, even disappears, with the mycelium underneath it. However large, to the truffler they are a cause for excitement as they show that the truffle's filament-like mycelium, which up until now has been living on the tree's roots (for up to eight years), has now spread out preparatory to fruiting.

Pierre let the dogs run and sniff and do what dogs do for a few minutes, then led them to *"un beau brûlé"* to begin the *cavage*, the seeking and unearthing of the truffle. He had swapped his nice shoes for Wellies, and donned a scuffed suede jacket with capacious pockets. He led the dogs to a burn, then let go their collars. *"Allez, allez! Cherche, Darius! Cherche, Octavius!"* Darius, the father of Octavius and at twelve years old a real pro at the truffle game, immediately set to dashing around in circles, nose to ground, back and forth, back and forth, and then settled in one spot. *"C'est où?"* Pierre asked. *"C'est où, la truffe?"* The dog moved between two spots two feet apart nose to the ground sniffing furiously, then returned to the same spot two or three times. *"Ici? Là? Montre-moi!"*

The dog dug once or twice then laid his paw on the ground as if to pat it, and looked up at Pierre, who took out his trowel and began to remove the earth carefully. When a treasure was unearthed, Pierre set it aside and, even before examining it, reached into his bag for a treat, congratulating the dog and rubbing his ears as he fed him the morsel. This happened over and over again as Pierre led the dog from *brûlé* to *brûlé*, with varying results. In quick succession, Pierre found a dime-sized *rufum*, a truffle but a variety of no commercial value. He found one the size of your little fingernail, also worthless.

If the truffle wasn't near the surface, he stopped digging every inch or two, thrusting his head into the dirt and sniffing deeply, picking up handfuls of the waste and sticking his nose into it. Curious, I, too, stuck my nose into the dirt in the spot where he had found a minuscule black disc and discovered that the damp earth reeked of truffle.

Now, there is no reek in the world like the reek of a truffle. Once sniffed, never forgotten. Cupping my palms around one of Pierre's larger finds, I inhaled deeply, eyes closed—and quickly drew away. Although a Frenchman would protest (we are squarely in the realm of acquired taste, or rather smell, here), I have to say that the perfume of a truffle to my perhaps unrefined nose is best compared to a mixture of four parts old gym socks, two parts boiled cabbage, and one part "Honey, did you leave the gas on?" Which is not to say it is unpleasant, merely unusual and very distinctive.

(And penetrating. All the old recipes for scrambled eggs with truffles instruct the cook to place a truffle together with the whole eggs in a closed container overnight. By the next day, the truffle has perfumed the eggs in their shell even before cooking. This adds to the intensity of the final product, a rich dish made in a double-boiler with usually *une noix*, or about a tablespoon, of butter to each egg as well as a dousing of cream.)

"There's one. A small one, which I crushed a bit." Pierre was saying. This was about the size of a marble, almost brown where the dirt still covered it, black underneath. The dog moved on to another spot, and Pierre suddenly moved in beside him, pulling Darius away. "This one is right at the surface almost. Hey! A field goal! Look at that, that must be ninety grams [three ounces]." And so it was. Large, nearly round, about the size of a large plum, the truffle gave off waves of that unmistakable, almost too-rich scent, the smell in fact of the inside of his Land Rover.

"This," said Pierre after searching for several minutes in yet another hole only to come up with a single black speck that he dubiously inspected in the light, "is the problem with dogs. Not only do they find black truffles, they find all the mushrooms that grow underground. And sometimes what you're looking for is minuscule, an immature truffle. Pigs—yes, some *trufficulteurs* still use them, especially here in the Lot—find only the ones that are ripe, but they're looking above all for themselves. They have more endurance than dogs, but they're harder to control."

Some of the truffles the dogs found that morning were only as big as a dime, some the wrong variety, some rotten or savaged by insects. The big one Darius had found almost right off remained the largest and finest specimen of that morning. Nearly palm-sized, it looked like an asteroid, with a curious dimpled volcanic/lunar appearance to the black skin. It was hard as rock, which also explains why you don't crush them underfoot when you traipse about the place. It was worth, depending on the market that day, anywhere from forty to fifty dollars, the recent price for a kilo, uncleaned, falling somewhere in the region of thirty-five hundred francs, about five hundred dollars.

"That is what is so exciting and mysterious about the world of the truffle," Pierre was saying as we returned to the car. "You arrive at the truffle ground, you see a new *brûlé*. Paaf! You find three good ones right off, of which one is one hundred grams! On the other hand, you go to another ground, which maybe has younger trees well-watered and cared for, and you find . . . nothing."

A half hour later we sat down at a long communal table in the Lion d'Or, a café/bistro in Lalbenque. It has a long bar on one side, with waiters coming and going, their arms full of plates of food and huge tureens of soup on their way to the long tables set with cheery checked tablecloths that take up the rest of the large room. Pierre got up to greet first one new arrival, then another. He nodded to people at several surrounding tables, identifying them to me, sotto voce, this one as a producer of truffle oil and truffle butter, that one a seller to restaurants, a third the head of the union of *trufficulteurs* of the area, over here the *courtiers*, or brokers, buying for wholesalers in other cities, over there a big charcutier whose company sold truffled pâtés and foie gras.

The talk all around us was of truffles, the weather, the price, the scarcity. ("Thirty-five hundred francs? Yaagh!" "There's nothing to buy, that's why, and the week before Christmas so the pressure's on." "Terrible weather. They all got burned in August, you know.")

Our table companions were an older couple in town to see the market, who, overhearing our conversation, interrupted to introduce themselves and ask a question of their own. Soon, Pierre's soup spoon was at rest and he had embarked on a recipe for *pommes de terres sarladaises à la truffe*, potatoes fried in goosefat, with diced truffles added for the last minutes of cooking. Conserved or fresh truffles? "I don't conserve them," he said, "because I like to eat each mushroom fresh in its season—morels in the spring, chanterelles in the summer, cèpes in the fall, and the truffle in the winter. And," he told them, "the black truffle is the only truffle worth eating raw, the Brumale,"—this is the second most common kind of truffle—"is too acid and sharp in flavor. But the black, shaved over a simple salad . . ." His eyebrows arched, and he gave a big smile.

A few minutes later, as we tucked into the *plat du jour*, rare *magrets* swimming in a deep brown Périgord sauce dotted with black islands of truffle, I asked Pierre the big why. Why are there fewer and fewer truffles every year even with the advances in cultivation they have made since they first began seriously studying truffles a generation ago?

"We understand a little but, but not all," he began. "We

thought when we invented the mycorrhized tree that we would see an increase, but we forgot that there are other kinds of truffles in the earth that contaminate the roots of the trees and compete with the melanosporum."

The mycorrhized tree to which he was referring is an innovation that appeared about thirty years ago. The *trufficulteur* can plant all the correct trees, in the right soil, that he wants. Yet, without the presence of truffles already in the ground as a source of mycelium, the roots of the newly planted trees will never be colonized and form the mycorrhizes that seem to serve as the reservoirs from which the mycelium migrates out into the soil each spring to fruit as truffles.

The peasants, of course, knew all about this a hundred years ago even if they couldn't have told you what a mycorrhize was. They used two more traditional methods, the first of which was simply to take seedlings growing around a truffle tree, in the wild or cultivated, already producing truffles, in the assumption that the offspring would share the same affinity for truffles as the parent. The second method involved baring the roots of a seedling and nurturing it for a time in a moist environment together with pieces of fresh truffle before planting it in a hole in which pieces of fresh truffle had also been sown. Today mycorrhized seedlings are produced by a semisecret, rigorously controlled, patented process involving sterilization of the roots and subsequent innoculation with melanosporum spores.

Scores of thousands of these seedlings have been sold over the years, and one would have expected to see, beginning about fifteen years ago when they should have begun producing, an upspike in annual truffle harvests. Which did not happen. Instead, said Pierre, "the last thirty years of cultivation have favored contamination. Now we're trying to limit that contamination." The guilty party is primarily the Brumale truffle, which has neither the taste nor the smell of the black truffle and which, besides, cannot be conserved. It is still sold, but at a much lower price, as is the Estivale, or summer truffle, which also competes with the black truffle.

While Pierre wasn't being intentionally disingenuous here, I would later hear (and usually quite vehemently) from many others in the truffle trade that the researchers had let their enthusiasm for early results run ahead of their actual knowledge. Specifically, they told people to plant hazelnut trees, which begin to produce at two to four years versus the oaks' ten years. Unfortunately, hectares and hectares of hazelnuts subsequently became contaminated with Brumale truffles, turning a costly investment worthless overnight.

It all gets both vague and complicated rather quickly, but it would appear that, in Pierre's view, the culprit of the modern era was the then newly arrived tractor, and more specifically the kind of intensive, yield-driven agriculture that went with it. Not only did farmers have less time for handwork, but they certainly weren't going to take a risk planting more truffle oaks when there were bank loans to repay every month (on the tractor). By cutting the brush not just around the trees but between rows and surrounding the truffle ground and by compacting the soil and turning it over, even at a very shallow depth, the cultivators destroyed what may have been a natural *cordon sanitaire*, a biological barrier keeping the competing truffles out.

This is a good example on the one hand of just how limited the scientific knowledge is and on the other of how fickle and demanding the truffle is. One of Pierre's ongoing projects involves extensive interviews with many of the older generation of *trufficulteurs* who have seen much better times. Over the last two decades at least, he has been collecting and attempting to codify what they know from long experience, and what their fathers and grandfathers had told them.

He and other researchers have then been evaluating these ideas more systematically but have been hampered by a lack of basic knowledge of the biology of truffles, even what they eat or how they fruit. And the necessarily long-term nature of these studies (twenty years being the producing lifespan of a truffle oak) has meant only incremental progress and maddeningly imprecise results. Debate still rages over such basic issues as whether one should even touch the ground (to remove brush or aerate the soil) around the trees of a *truffière* that has begun producing or to prune their branches or remove the seedlings.

While we had been eating, just across the street from the restaurant the stage was being set for the weekly winter drama known as the Lalbenque Marché aux Truffes, the reason this restaurant was packed on a midwinter Tuesday. "Just wait and see how much green you see today," Pierre had said, with a crafty smile, on the way over. It took me a few moments before I remembered that green is the color of the five-hundred-franc note, the French equivalent of the hundred-dollar bill.

By the time we left the restaurant after our meal, the main street had been transformed: the opposite sidewalk had sprouted a line two hundred feet long of low wooden tables, with a blue rope barrier set up along the street in front of them to keep people back. As it was almost two P.M., the tables, and the streets, were full. Behind the tables against the far side of shops, all closed, a lone red flag hung limply from a standard in front of Groupama Assurances. A few doors down from the restaurant, on the stone steps leading to the massive wooden doors of the Lalbenque town hall, a sturdy table had appeared, on it a large lidded box complete with conspicuous padlock.

Yet the first thing to strike me was not the crowd or the preparations, but the scented breeze, for the whole town smelled of truffles. Behind the tables, more than fifty sellers stood elbow to elbow, smoking, chatting, looking out at the crowd for the faces of the buyers, who would not approach until the market was officially open. These were almost entirely older faces, the *trufficulteurs*, men in ancient leather jackets with red noses shining under their wool caps and women with bright hair clad in everything from muddy farm jackets to belted raincoats and bulky sweaters over their housecoats.

In front of each were their wares, held in a remarkable variety of containers, delicate woven reed baskets that might hold a handful of coins at a pinch to enormous picnic panniers, a plastic bag from the local supermarket with the edges turned down, a tiny paper sack, even, that might have once held penny candy. The contents were well hidden, most by dishtowels or canvas cloths, and would remain so by longstanding tradition until two-thirty, whether to add drama to the proceedings or to keep potential buyers from having an unfair advantage I didn't know. Why two-thirty? The gendarme looked at me oddly when I asked. "One must have lunch, of course," he said simply.

I walked up to the last seller in the long line, a substantial character with a hand-rolled cigarette screwed into the corner of his mouth and a dour look on his unshaven face. Was it going well, the season? I asked. Pfffft! he began, letting out his breath in a hopeless sigh. He gestured at the brown paper sack in front of him, then, looking out at the crowd, casually unrolled the flap and let me have a quick peek at the three or four walnut-sized truffles nestled in the bottom. "I have only two hundred grams to sell today. No, it's not good. This summer, the rain—maigre, maigre, not enough, that's all, and then no frost yet, either. Because that's when they really grow, you know, the month of December after a good freeze. The price? Who knows . . . three thousand, thirty-five hundred francs? How do you judge the quality? You have to be an expert, by the smell, mostly."

Precisely at two-thirty, the red flag fell and two blasts of a whistle sounded, the signal for a smartly clad gendarme to drop the blue rope. A dozen buyers rushed forward, most making a beeline for one seller or another, at the same time as up and down the line the cloths came off the baskets. I saw a pathetic offering of three tiny black marbles right next to a basket a foot and a half square overflowing with tubers the size of tennis balls. Most people, however, just did not have that much to sell, a couple of handfuls, perhaps a pound or two being the norm.

Everywhere I looked, the same ritual was being enacted, a buyer approaching a seller, peering at the offering, leaning down, inhaling deeply, perhaps, then, at a nod from the seller, reaching for a single truffle to gauge the quality, finally scribbling a price on a slip of paper and passing it across the table. There followed either a handshake or a shake of the head. In the former case, the seller covered his or her basket, and buyer and seller promptly disappeared. If the seller refused the offer, he waited until another buyer approached, whereupon the same ritual was repeated, but with the risk that, given that the market lasted only a half an hour with so few truffles for sale, the price offered would only decline. Nothing prevented that first buyer from returning with a lower price, either.

That would seem to be it, all except for the weighing. Some pairs of buyers and sellers indeed could be seen together across the street at city hall, where the padlock and lid had been removed from what proved to be a scale. After the weight was agreed on, green bills changed hands in large quantities (the whiskered fellow's two hundred grams would have been worth one hundred dollars that day), and both parties generally repaired to one of the two bars on the street for a drink.

On the side streets, however, I saw a different scene, the whispered words passing between buyer and seller having been more like "My car is the green Peugeot wagon behind the post office." Here, men and women gathered around open trunks of cars, one usually holding a scale from which hung a wire basket full of truffles.

In theory, the price offered is for the whole basket, with the seller not having the opportunity to sort through the merchandise or shake off the excess dirt that clings to the truffles (up to fifteen percent of their weight, before cleaning and brushing). The basket was thus a sort of control, Pierre told me. "The wire basket lets the dirt fall through," he explained. "And the buyer sees if there are rotten truffles. 'I don't want them,' he can say. 'Put them aside.' *On peut marchander.*" They can bargain. "And they discuss it. Maybe he lowers the price. Maybe he separates those he doesn't want. But in the wire basket, which is particular to this market by the way, you can see everything—those that are rotten, those that are eaten up by insects."

And the total amount sold at the market that day, I asked. "We can check after, but I think you'll find it was somewhere around sixty kilos," Pierre said, one of the worst weeks in living memory. It was indeed going to be a very bad year for truffles.

THE TRUFFLE MERCHANT

In his dimly lit office the day after Christmas, Jacques Pébeyre was telling a story. In the rather drab, 1940s surroundings of dun-colored ersatz wood paneling and black-and-white truffle-hunting prints, with a single door at one end of the narrow room, a cluttered desk and chair, and behind that a pebbled glass window letting in the gloom of the rainy afternoon, the story was apt, long and rumpled and sad, and told by a man who was nearly of an age to have seen it all, or at least the worst parts. This is how it went.

The last years of the nineteenth century were a terrible time for the Lot, as they were for every grape-growing region in France. Phylloxera, the parisitical insect infestation that attacks the roots of the vines, laid waste to vineyards, destroying over three decades the work of hundreds of years, ruining farmers, barrelmakers, ox and barge transport companies, indeed anyone connected with wine. The blow was all the more severe in this particular region because the chalky, pebbly soil and rough terrain of the Lot lent itself particularly to grape growing as it did to nothing else, even if it had meant terracing the steep hillsides and working most vines at best with oxen, at worst completely by hand.

The wine from the grapes was the only commodity the Lotois had, the only product they exported aside from small quantities of livestock, pine resin, and walnuts. The wine was thus the only more or less reliable source of cash, the economic lifeblood of most every village and town. The long and short of it was that nothing else would grow on the arid, limestone hills except spindly oak, and weedy grasses that give the *causse* its almost unchanging palette of tans and browns and the rare clump of ragged deeply green juniper or *buis*, wild boxwood.

In the midst of disaster, however, the farmers began to notice that the land was changing. The hillside vineyards, either stripped or abandoned to the dead and dying vines, began to grow over with scrub, especially two ubiquitous varieties of tough dwarf oak. On other barren slopes, the government began to subsidize their planting to battle erosion and as a sop to the farmers. And around these trees *brûlés* began to appear, telling the farmers that something was happening under the ground. When fall turned to winter, they began to find, in increasing quantities, black truffles.

Soon they began to plant the oaks intentionally, in plantations, where there had once been only vines. While the existence of wild truffles in the forests of the Lot had been known to the peasants at least since the sixteenth century (it was a winter vegetable, after all) and most likely long before, the idea that one could, in a sense, cultivate them by cultivating the trees around whose roots truffles seemed to live was a new one.

Abundant truffles, a healthy demand in northern cities, and newly completed railroads to ship the delicacy fresh overnight to Paris and other major European capitals; it seemed an auspicious start to the new century. The two cusp decades, however, would come to be seen in later years as the golden age of the black truffle, the harvests declining decade by decade, at first a little and then precipitously. The generally accepted figure for French truffle production in the year 1900, for example, is one thousand tons, a staggering figure given paltry modern harvests. Today, Pébeyre finished, "there are fewer and fewer truffles."

He paused, regarding me over the expanse of desk between us, his hands toying with a pen, and the phone rang. He picked up the phone in front of him, put it down, reached for a portable phone beside it, and pushed one button after another, holding the phone to his ear each time as the ringing continued. "Which line?" he finally called to his secretary. "Which phone? I hate these things," he muttered, before launching into a lively conversation with someone on the other end about a recent party and how fine a fellow the English ambassador was. Pébeyre is a solid man, with ruffles of white hair ridging his completely bald crown, and large, almost square ears. He has an equally generous, almost perfectly pyramidal nose above a strong chin, and green eyes set deeply under a much-wrinkled brow. But it is the hands that captured my eye, the fingers gnarled and weatherbeaten, the hands of someone who had lugged and washed and sorted his share of truffles in the years before he turned over the everyday running of the business to his son.

"I have worked with truffles since 1946," he said after he hung up the phone. "More than fifty years! And I grew up with my father, who worked eighty years with truffles, and my grandfather before him. Me, here, Tuesday the first of February, 1949, there was the market in Cahors in the morning. There were four tons of truffles sold. And in the afternoon, there was the truffle market in Lalbenque—also four tons of truffles sold. I myself bought fifteen hundred kilos at the markets for our business here. In one day back then, that eight tons represents one-half of the total yearly harvest of the whole of France today, between sixteen and twenty tons."

Jacques Pébeyre is a businessman, and his only business is truffles. He comes from a great and storied Lotois family who were *trufficulteurs* long before they turned to preserving and wholesaling truffles as a more reliable venture than growing them. He is a man of some stature in the Lot, a man with, quite literally, the dirt of the sorting table still on his hands. And yet he is also quite worldly, his name known to the best chefs, restaurateurs, and hoteliers around the planet, and a man as likely to have just gotten off the plane from a business trip to Hong Kong, Singapore, Tokyo, and Sydney as from Paris or London. His long, intimate association with the fickle mushroom gives his words an authority others lack, a perspective broader and more incisive (and critical) than either the researchers or the growers.

"There are several reasons why there are fewer truffles," he began. "The first reason is a human one. There is no longer anyone living in the countryside! No one to plant the trees and to take care of them. You had maybe seven or eight people living on every farm back then, and then in the next generation there were only two, the old couple, and today no one." In less than two generations, the population of the Lot has shrunk from three hundred thousand to one hundred fifty thousand people, with the majority of those living in towns and villages. "The old peasants," he continued, "they knew what to do, when to do it, how to do it. They didn't know why. But they knew when to prune their trees, what the ground needed, because they observed, they remembered, and they passed on their memories. It was the time when there was a deep relationship between nature and man.

"And then there was Verdun. It was largely the farm which provided the sons who went off to fight. And half of them didn't come back. The half that did found the land in bad shape, neglected. The two great *étapes* in the decline of the truffle here in the southwest are in fact 1914 and 1940, corresponding to the two wars. Beginning in 1950 it was steeply downhill, with eighty to one hundred tons the average, then by the 1970s, thirty to forty tons. Today, we have arrived at a moment when the least atmospheric disturbance is a catastrophe for the harvest."

While the usual suspects of the modern age—industrial agriculture, pollution, mass migration to cities, and the erosion of traditional values—featured in Pébeyre's catalog of reasons, he reserved special ire for the *scientifiques*. "My *father*," he told me adamantly, "wrote letters to the scientists at the agricultural colleges and research stations. I still have the carbons of them. He wrote to sound the alarm about the fall in truffle production. And they never paid attention. That was fifty years ago!"

I had been told that truffle trees were being planted at somewhere around five hundred to one thousand acres a year over the last decades. Yet when I mentioned this, he scoffed. "The people who have been doing the replantings recently, they're retired, or incomers who don't know how to take care of the trees! *Trufficulture* is an art, and when those who practice the art, those for whom it is truly a part of their life, when they disappear, the truffle disappears. I am just as guilty as others," he continued. "I myself, I planted twenty hectares of truffle oaks fifteen years ago. And the results have been pathetic! I don't take care of them enough. I blame myself. You have to be there. To work them with your own hands, the trees. Prune them, walk among them and see how they're doing, and often. I haven't been to my trees yet once this season, even to harvest what truffles there may be! But," he said with a laugh, "I'm pretty sure someone else is taking care of that for me!" (Nocturnal truffle thievery, once the inspiration of much local lore and, sometimes, violence, but now less prevalent, seems to have been one of the great hivernal pastimes in some corners of rural France.)

Yes, the truffle is mysterious and capricious and something whose demands require total devotion and generations of hard work to master, but the less poetic side of the coin was its value to people who never before had seen such money. "The huge abandoned farms you see out in the country, in the middle of nowhere, the big houses and vast barns, they were built on truffles. For the truffle was part of the riches of this particular countryside, a little money which fell from the sky each year on top of what they earned from the harvest and the beasts. It was cash money that the peasants didn't have to spend on *les charges* [their everyday expenses and bills] at the end of the month," Pébeyre said before launching into a story about a friend's father's ledgers, his written record of how many truffles he harvested from where and for how much he sold them at what market going back to the 1890s. "Around 1900," he finished, "it took only two years of truffle harvests to build a huge barn in stone thirty meters on a side. That was when there were the weekly farm fairs at Catus and Salviac, when you would see three hundred to four hundred kilos of truffles each week at the fair at Cahors."

Still, I told him, it was hard to believe there were no solutions in the modern era, given all our advances in agriculture and the sheer progress of science in one hundred years. His face went through several contortions as I spoke, from a moue of disappointment to a frown and ending with an outright scowl.

"In the 1970s and 1980s," he responded. "the farmers had gone to agricultural college. They learned how to raise cereal crops, or sheep, or beef, or chickens. They weren't interested in *trufficulture,* and, anyway, now they live on government subsidies. Then, the largest replantings of truffle oaks were done for the reason of subsidies; the government providing free trees, money for clearing the land of brush, for preparing it . . .

"*Les scientifiques,* they can do good, but for a long time they would have us plant hazelnut trees instead of truffle oaks. And now all those truffle grounds are infested with the low quality Brumale variety of truffle and not the black truffle. Why did the scientists tell us to plant hazel trees? Because they grow faster! With a truffle oak, it takes ten years until they're six feet high. With a hazelnut, two years! The scientists didn't think it mattered what kind of tree you planted [but] they were wrong! Sourzat's results have not been good, and the problem of the Brumale keeps him up at nights!" he finished, seeming to hint that this was as it should be, since Sourzat and his colleagues were among the aforementioned *scientifiques.*

"And there is another thing," he continued. "They always said the truffle oak grows best on the worst ground, dead ground, poor ground. What they were looking at, a lot of it, was vineyards grown over in truffle oak. Now this is my personal opinion, but what they ignored was the fact that the peasants used to fertilize the vines with *la colombine,* the manure from pigeon droppings, which is very rich. And they fed the vines every year so that some of that fertilizer stayed in the soil, enriching it. Of course, over the years, this natural fertilizer was used up, and . . . *hein?*" He threw his hands in the air and raised his eyebrows, leaving the conclusion to me.

He shrugged, a bittersweet smile flitting across the timeworn features. "There are subsidies, research supports, which attract the researchers. You can't blame them for taking money to keep their laboratories going; they have salaries to pay . . ."

Out in the main room of the workshop a few moments later, it was hard to believe that there was anything amiss in the world of truffles. In the center of the room were two long rectangular tables whose surface had been squared off into a dozen sorting bins by wooden dividers. Here and there lay funny-looking shovels with handles only a foot long, whisk brooms and dustpans for clearing away the dirt that clung to the truffles. The air was heavily freighted with their smell, and in one corner an apron-clad young woman weighed and stamped made-up vacuum-sealed parcels while across the room Pierre-Jean, Jacques' son, sorted and vacuum-packed other orders from a four-foot-square tabletop covered six inches deep in darkly gleaming, freshly washed truffles.

Off to the left was a small room mostly taken up by the cleaning machine. As we walked in, Jacques said, "Normally we would be cleaning three hundred kilos a day. These days we're doing only fifty kilos. At the truffle markets, usually we pay eighteen hundred to two thousand francs a kilo. This year—thirty five hundred minimum to forty-five hundred francs. But when you buy a kilo, fifteen percent of the weight is dirt, and then we might have to sell half of them as *brisures*, and there will be some Brumales and not black truffles . . ." *Brisures*, the lowest quality, aren't even whole truffles, but the larger pieces of truffle skin and chunks that have fallen off in the cleaning process. "Also, lots of the people selling truffles at the market have already taken the best elsewhere. Maybe a restaurateur, you know, who has already picked over the basket, taking the best off the top. When I buy a basket, I want to buy the most beautiful and the less beautiful, the whole basket."

From November to March, truffle season, his *courtiers*, or buyers, go to all the truffle markets in France, not only Lalbenque, but Martel, Richeranches, Valreyas, Carpentras, Aptes, LeCan, and to Spain as well, although this year a drought there has meant a precipitous decline. "Truffles, fresh and conserved, are our only business . . . although we also make and sell a truffle butter and truffle-scented oil. We sell three times as much conserved truffle as we do fresh truffle . . . We work the year round, though it's true that July and August tend to be the quieter months."

Pébeyre *et fils* buy the truffles *en terre*—unwashed and unsorted, by the basket. One kilo unsorted may consist of up to fifty percent of the lowest grade only, a mixture of Brumales and black truffles whose quality is diminished by rot or the ravages of insects. When conserved, canned under pressure in an ancient autoclave the size of a Volkswagen in Jacques' back room, the truffles also will give up about twenty-five percent of their weight as juice, itself a valuable commodity. All of this explains why a pound of glistening premium grade, perfect, round truffles sells for between eight hundred and one thousand dollars in Paris at the retail level.

In the United States, Williams & Sonoma was recently offering single conserved truffles of about an ounce for fifty-four dollars a jar, or eight hundred sixty-four dollars a pound. But, says Pébeyre, buyer beware, especially of truffles served in American restaurants. "Eighty percent of the truffles sold in the U.S. are not good, not even the right mushroom, not tuber melanosporum but rather competing species like Chinese truffles or summer truffles (*l'estivum*), which ripen in June and July and which have neither the taste, smell, nor the color of the black truffle, being whitish to marron on the interior. La Brumale is also a winter truffle, very close to the *noire* in aroma but with a flavor less powerful, more bland and insipid. So what is sold is not the real truffle, but those lesser things that are usurping the good name of the true truffle."

From the sorting table where he was cutting away the rotten parts of still-wet truffles, Pierre-Jean, who had been following our conversation, began to speak. More compact and intense than his father, Pierre-Jean spoke with all the intensity and even anger of someone who is slowly watching his livelihood disappear before his eyes.

"Truffles," he said bitterly, "are in a free fall, a dramatic situation, but these stories you see in the media make it seem like everything is just fine. I am younger than my father, and in the last years I have begun to realize that most people value the image of the truffle more than the product."

"Around the truffle there is a whole industry," his father added. "And the people who are doing the mythologizing, they make a living not from truffles, but from all the research into them."

Or worse, his son continued, "they are making a living from the image, the pretty picture, which is not at all pretty when seen up close, as we see it here, where the real harvest, actual truffles, is our only business. Those who do it—who plant the trees and work the land and take care of it and then gather the truffles, they don't talk about it. But there are a lot of others on the edges. They tell stories, they are the ones participating in the creation of this myth, this legend, rather than doing something about the problems."

He sighed deeply and put the knife down, wiping his hands on his apron, and turned to engage me fully.

"There are truffle journals and truffle societies and truffle Masses in the churches and a brotherhood of the truffle! Even a patron saint of truffles—St. Antoine—who never was until four or five years ago. They have the image, all the pretty pictures. And not the thing. Because that takes work, while the image is easy. The poetry of the truffle is in the mouth, on the plate. If that disappeared, that, that would be sad. At Lalbenque [truffle market] the other day there were fifty kilos of truffles with five hundred people staring at them. That's pathetic. And yet most people are satisfied with that kind of media image . . .

"As for the scientists, well, they blew it thirty years ago. But—"

"*Toutes les vérities ne sont pas bonnes à dire.*" His father interrupted. Some truths are better left unspoken. "Sourzat's mycorrized trees?" he snickered. "Twenty-five years ago we planted six hundred trees on twelve hectares. It was us, and a nurseryman, a researcher, and a foie gras producer who all got together. It was a big experiment, best advice, best trees." He rubbed thumb and forefinger together to signify that it had also been quite a costly endeavor. "A big experiment, and a big failure."

"We've been planting them in France for thirty years now," Pierre-Jean continued bitterly, "without knowing the most basic things like when little truffles are formed and what they eat! In the sixties France produced eighty tons of truffles a year. Since then, divide that by four. When you think of this terribly expensive product which comes from an otherwise poor region, that's a large failure. But still we planted those trees, hundreds of thousands from the seventies to today . . . Can you imagine selling for thirty years apple trees which produce no apples? That's a scandal. But now we ask ourselves if maybe the point wasn't to sell the trees . . . In the seventies, okay, we made a mistake. But to make the same mistake over and over again for thirty years?

"And who bought them? They were sold in regions that have never produced truffles!" he continued, his voice rising. One of the women working the scales looked over at his vehemence, giving me a smile that told me she'd heard it all before. "They were sold to people who have never produced anything, people who have a business card which says '*Trufficulteur*' which goes over big at some party in Paris. These people, maybe they produce truffles, some, but if they're only enough to eat themselves, that doesn't interest us. So the whole thing is scandalous, in my opinion.

"The problem is a simple one: we need to do the basic research to understand truffles and then we have to find the people to pass that knowledge on to. Because the rural exodus continues. On the land between Lalbenque and Limogne the population density is four people per square kilometer, which is what you'd expect to find in a desert region. If the truffle is going to survive, it will have to be some kind of intensive agriculture because there are no men left on the land. And the ones who leave are the smartest . . ."

Almost two months later, seventy-five-year-old Pébeyre Sr. was on his way home from the Wednesday truffle market in Richeranches, in Provence, when his car was forced off the road and into a field by a big BMW. No sooner had he gotten back on the road when another car pulled in front, blocking his passage. Six thieves piled out and, while Pébeyre's wife watched in horror, forced him out of the car. They made him open his trunk, then fled with 150 pounds of truffles worth thirty-eight thousand dollars.

Such violence over some black mushrooms? I still didn't quite understand what all the fuss was about.

THE CHEF

A week later, Jacques Ratier called. Did I want to come up to the restaurant at the end of the month? I thought you were closed until March, I told him. This is special, he continued, explaining that they were opening for one night. A group of Americans from California was coming through with Babé Pébeyre (Pierre-Jean's wife) who had been guiding them around France, particularly the Lot, on a truffle tour. Which is? "Well, me, I'm making them a whole menu, truffles in every course," he finished. "They started at Ducasse, in Paris, a truffle-tasting menu, then came south, eating truffles every day and every way on the road down."

When I walked into the restaurant, the usual organized chaos reigned, with a few exceptions. Noëlle, who was flitting back and forth from kitchen to dining room in a white cashmere turtleneck and long narrow black wool skirt, seemed dressed more to eat dinner than to serve it. She was laying out glasses on a single long table for fourteen in the otherwise empty space. On a smaller table to one side, the wine was displayed, a three-liter jereboam of Château Lagrezette 1997 and eight bottles of Château Eugénie Cuvée Reserve des Tsars 1995, two excellent *vins de Cahors* that would give up their corks one by one as the evening advanced.

When I raised my eyebrows at her attire, and the intimate room, Noëlle said, "I'm going to eat with them, the group." I raised my eyebrows again. "The truffle tour is meant to be very special, a small group which pays an enormous amount of money for the opportunity to go places, meet interesting people, and eat truffles at every meal, all things you could never do anywhere else," she said. "They chose us for this stop, I think, because this is a very homey place, simpler food, *plus relaxe*, a change from the Michelin circuit they've been on. And I speak English!"

In the kitchen, Jeannette, Jacques' mom, was washing lettuce at the dish sink, singing along with a radio turned low, an apron tied carefully over her good dress, as she would be waiting table. At the stove, her son sautéed *bacon* (unlike American bacon, this is more like a dry-cured *jambon de Bayonne* or a prosciutto) on one burner and seared generous fillets of pink-fleshed perch on another. He was unusually quiet, moving rapidly from task to task, obviously under pressure for an occasion of some importance. He was a bit ennervated, but also excited, his glance straying every so often to a dishcloth-covered container from which emanated the most interesting smell. It was the first time he had been asked to cook for this very exclusive event, and tonight's group had some serious eating behind them, Le Gindreau and Le Balandre, two Michelin one-stars they had already visited locally, not to mention Ducasse's three-star restaurant at the Plaza-Athénée Hotel in Paris.

"Michael, could you . . . ?" Jacques asked, ending his sentence with a whistle and darting his eyes to the back room, a message I had by now come to understand quite well. I went and got a white apron and washed my hands. "Jeannette is serving tonight, and Noëlle is eating with them, so I could use a hand with the plating," he explained, not looking up from the flashing blade of the tiny truffle mandoline that shaved ultra-thin slices from the truffle in his hand, sending the first waves of its heavy perfume into the kitchen. Next to the machine sat the square glass casserole dish that he had been eyeing. I leaned over and lifted the cloth, inhaling deeply of the black treasure it contained. This was going to be some night, I thought, and I didn't even know the menu.

Jacques caught the smile and grinned back like an eager adolescent. "*Oui, oui!* It's not every day I get to cook with these." He lifted the cloth off completely and ran his fingers over the rumpled black surface of the truffles, almost caressing them. "Just wait. There's nothing like it in the world." He was preparing six courses, not including the *amuse-bouche* that would start them off, and I was interested to see what he had come up with.

The group arrived in four cars about eight o'clock, fourteen people and a black Labrador tumbling in out of the cold February

rain. I watched from the door of the pantry, where I had been cutting bread, as the *bonsoirs!* and *enchantés!* rang out. The bar filled
and became instantly raucous with the clink of kir glasses and
merry Californians in evidently high spirits, and I had a sudden
unexpected pang of homesickness. It struck me that this was more
Americans in one place than I had seen in more than six months. I
listened to them talk, the lovely, mellow accent striking my ear
pleasantly, such a contrast to the clipped tones of our English
friends here.

Jacques had gone out to welcome the group, mingling with an
untouched glass for ten minutes as he was introduced to this one
and that by the two he already knew. One was Babé Pébeyre,
daughter-in-law of the wholesaler who sold Jacques his truffles.
She was tiny and lively, her short stylish blond hair swinging as
she moved easily in the crowd, helping everyone feel at home,
punctuating her French-accented English with waves of a black
cigarillo. I saw Ken Hom, too, along for the ride as a special guest
and local expert. This Chicago-born Chinese-American chef is a
household name in Britain, and in many European and Asian
countries, as the star of BBC-produced cooking specials as well as
the author of a score of well-received, best-selling cookbooks. He
has a summer home in the Lot and comes to La Récré as often as
he can.

As Noëlle moved them all to the dining room, Jacques came
back to the kitchen and finished putting the last touches on the
amuse-bouches. This hors d'oeuvre—*cocos truffés au homard*—
was an exquisite six bites of fresh white bean salad dotted with bits
of lobster and truffle and crowned by an entire claw, the whole
enclosed in a ring of cucumber slices. As Jeannette served them, he
was at the stove again, giving the soup a last whisk, then laying out
fourteen bowls on the pass, and shaving, shaving, shaving truffles.
"You know who's out there?" he asked.

"A group of hungry Americans?"

"Yes, but there are also two chefs. One is from Le Prunier in
Paris, a two-star. The other is French, too, but he lives in

California. He's on leave or something. From a restaurant in Berkeley? Chez, chez . . . I can't remember what Ken told me."

"Chez Panisse?" I asked.

"Yeah, that's it."

"Jesus Christ, Jacques!" I told him. "That's the most famous restaurant in America. Have you ever heard of Alice Waters?" He shook his head, and I went on to explain who she was and the revolution she had led. Of course, to a Frenchman, that such notions as eating each food in its season, cultivating local farmers, and the benefits of organic meat and produce could be revolutionary is itself revolutionary, showing how differently our food cultures have evolved.

As we crumbled bits of truffle into a rich velouté of leek and potato and floated whole slices in a blossom pattern on the top of each bowl of soup, Jacques couldn't resist gathering up the crumbs from the counter and tossing them into his mouth. "They're good just like that?" I asked.

"You've never had truffles raw?" I shook my head, and he held out a cutting board black with slices. "Truffles are mushrooms, incredible in a salad, not just in a sauce, which is the only way most people eat them now." I crunched down on a whole slice (they have a texture somewhat like half-cooked risotto rice), my mouth filling with an intense musky explosion for which I had no comparisons, except to smile. By then, Jeannette had cleared the hors d'oeuvre plates, and we both began ferrying the bowls of steaming soup out to the bar, from where she could more quickly serve the guests.

When I came back into the kitchen, I found a smaller bowl on the pass, and looked at Jacques, who was grinning at me, holding out a spoon. The soup was thick with truffles just warmed by the heat of the soup. "This is the first time in my life," I said, "I've really been able to taste them, you know." He had been watching my face. "I know. I know," he replied.

Noëlle stuck her head in the kitchen, looked right at Jacques, and said, "Ken salted the soup." He didn't respond, and she said, again, a touch sarcastically, "Just so you know, *Ken salted the soup.*" He shrugged, lifted a ladle of soup, tasted, shrugged again. The deeper significance of this eluded me, and I asked what was going on.

"It's an error," Jacques explained. "The plate should arrive at the table perfectly seasoned, perfectly salted and peppered. What can you do?" He sighed and went back to work.

The entrée was his famous lobster ravioli in coral sauce black with bits of truffle and generously garnished with slices almost hiding the pasta. After we had plated it and helped Jeannette serve, I again came back to find a saucer with a few ravioli on it, and more truffles. "You don't cook truffles." Jacques was saying as I took my time with the offering, cutting each ravioli in half and bathing it in the sauce. "They say it starts to release its maximum aroma at about ninety degrees. The truffle is like that. You put it in the dish raw, the heat of the plate makes the flavor jump out in your mouth. But if it gets *too* hot, you kill the flavor. And it is very rich, the truffle, which is why I didn't finish the coral sauce with butter as I usually would. It would have been too heavy."

While the noise from the next room increased in proportion to the sinking level of the wine in the bottles, Jacques continued his lesson. "And their flavor marries well with fish and seafood, especially those with finer flesh and a lighter taste, like the perch they're going to eat next." The whole time he had been moving from task to task, taking perfect rounds of potatoes Anna that he had prepared earlier and sliding them into the oven to heat along with the perch, for its final two minutes of cooking.

"There are wines that smell like truffles, that smell of the earth, of dirt," he said a little later, his nose plunged into the dish of truffle shavings. They were for the fourth course, a simple green salad dressed with a truffle-laced vinaigrette.

The plates of perch were beautiful by the time we finished putting them together. The golden rings of potatoes went first, their edges peeping out from under a square of pink perch seared crispy brown on one side, a slice of the dark bacon, a single petal of peeled, confited tomato, the whole overflowing with slices of truffle and surrounded by a moat of veal stock thickened and enhanced with more truffle. Sadly, there was only enough perch for the guests, and I had to content myself with a bit of leftover potatoes Anna—covered in truffle crumbs—that Jacques and I shared. "Mmmmm," he groaned. "*Merde* but that's good!"

And so it went, through a cheese platter and a very light *gratin de fruits rouges* for dessert. We joined the big table for this last, in time to catch the by-now very animated crowd in spontaneous performance, and apparently having just missed Ken and Noëlle perform a song. The group had indeed worked their way through a bottle or six, and each consumed a king's ransom of truffle in the course of the meal. Little wonder they were so happy.

NINE

RITES OF THE GRAY DAYS

In our mailbox the day after Christmas there appeared a half sheet of paper on which was written: "On the occasion of the New Year, the mayor and his municipal councillors invite you to gather to raise a glass to friendship at the town hall on Sunday, January 7th, at 2:30. (signed) Monsieur le Maire Raymond Laval."

On the appointed day, Lily and I made our way up the path from Auricoste, coming out to find the village square filled with cars and people, some of whom we hadn't seen for months, and many we had never seen at all. As always, it was raining, and we hurried into the large room beside the mayor's office to find it filled with people and fragrant with the smell of wet wool and cig-arette smoke. The crowd was mostly older women, some couples in their fifties and sixties, a few handfuls of people our own age and their children. Around the walls was a single row of chairs, occupied almost entirely by the very old, all in their Sunday best and the women having recently been to the beauty parlor, judging from the brilliance of their hair color and the perfection of their coiffes.

The Laval clan was out in full force, the mayor and his wife, their son, Alain, with his wife, Michelle and their two boys, Melvin and Yann, these last attracting Lily instantly like a magnet. A clutch of councillors was making the rounds, too: the lanky, ruddy-faced farmer Bernard Bousquet with his wife, Jeannine, and their shy daughter Valérie; sturdy Madame Tanis whose farm I often jogged past up at Fazende. Sylvie Lacombe, an earthy, voluble woman as often glimpsed in overalls forking hay to her husband's cows as in dress and pumps lending a hand at every village function, stood talking to Simone Bladié, thin and sere and solemn and very hard to imagine as jolly Madame Delfort's middle-aged daughter. Monsieur Laubin, Yves Astorg, Vielcazal the younger, and Gérard Laval, Raymond's eldest son, were also there, familiar faces and important today in their role as councillors handing around the refreshments while doing discreet lobbying for the upcoming elections.

Jacques and Noëlle showed up, too, as they did at every event of any significance in the village. While they were there to be neighborly, Jacques was running for a seat on the council, and it was important for him not only to show his face, but to get to know a little better some of those he had only glimpsed from the kitchen of the restaurant. Although I didn't know it at the time, he was a little bit anxious that he would not be elected. In this setting, it was true, he seemed a bit out of place, a nervous bundle of energy not quite blending in, slipping out for a smoke when he couldn't huddle in a corner any longer with his great friend, Bernard, there to talk about hunting, mushrooms, and their plans for the restaurant garden in the spring.

I caught sight of Patrick Cantagrel, moving in the background from group to group, shaking hands and doling out kisses as befitted someone readying himself to step into the mayor's sabots. Patrick, a stocky man in his early forties with a full head of dark hair and prematurely graying mustache and beard, exudes the confidence of one who knows what he wants but also has a gentleness of manner that suggests he won't tread on others to get it. Always seen in good suits of sober colors, and with a handsome, open face, he is a person the older generation seems to find reassuring, someone who went away to make his career, but then came home to take his up his duties after the recent passing of his father.

I waded in, hailing those I knew and introducing myself to others, making excuses for my wife who had stayed behind with a headache. Some asked how my book was going, if I hadn't had enough of them all yet. These were the most delicate of probes, respecting the country way of never seeming to inquire directly or deeply into someone else's business while letting you know they are aware of where you've been and to whom you've talked. As I chatted, I caught glimpses of Lily scurrying around, here on a lap, there giving *bisous* to a beaming old lady, who obviously took great pleasure from those little kisses.

People clustered in small groups wishing each other *Santé!*, good health, and *Bonne Année*, Happy New Year. The middle of the room was taken up by an enormous refectory table covered with bottles of sparkling wine, orange juice, soda, and rounds and rounds of a single kind of cake, *galette des rois*. The door constantly opened and closed on new arrivals, as around the room condolences were exchanged for recent losses, grandchildren were celebrated, and friends too old to get out much found themselves once again in one anothers' company.

After more than half an hour, with the room filled almost to bursting and the air hot and heavy from the gathered bodies, the councillors finally began to pop the corks and hand around glasses of lukewarm sparkling wine. The room fell suddenly quiet, but nobody drank, all eyes focused on Monsieur le Maire, Raymond's dark eyes drifting here and there as he first cleared his throat and then began his speech of New Year's welcome. This was, he told us in a few halting, heartfelt sentences, his twenty-seventh, and last, speech as mayor, for the March elections would return someone else to take his place (here he nodded, smiling, at Patrick), as he would not stand. He reminded us of the long way the village had come in that time, pointing out the thriving Zadkine Museum, the well-advanced renovations on the presbytery and five stone houses across the square from us, a project that had been ten years in the making. There, despite recent construction delays and setbacks (the lead mason had died unexpectedly), the expanded Les

Ateliers des Arques would soon, he hoped, be welcoming another group of visiting artists-in-residence who would be able to live and work throughout the year in their new quarters instead of simply staying for the summer as in the past. It was one more step in restoring *le patrimoine bâti,* the old stone structures, of the village, which, along with the new sewer system and the burying of the electric and telephone lines, was part and parcel of the council's vision of a Les Arques reborn and remade as *un pôle culturel*, a center of cultural tourism in the rural idiom. All of this he said, but in far simpler words.

People murmured their affirmation all around, and he passed to other things. "I'm so happy to see people from all of the different villages of the commune here. A good turnout, even in the bad weather, is important, I think, because we don't very often have the time to get together like this." By the villages, he meant the fifteen or so hamlets, gatherings of two or three or sometimes more houses tucked into the hilltops and hollows of the commune and where, in fact, most of the residents of the commune lived. I had always found that, when I asked locals where they came from, they never said "Les Arques." It was always the *lieu-dit* of the hamlet, Thémines, Les Granges, Les Places, Le Truc, Lesquirol, Bout, Le Camus, Lerissou, La Combette Sarrau, Fazende, to name just the most important.

He went on to introduce newcomers, relative or not so, the Dézys who, after visiting for years, had finally moved in permanently, Peter and Cathy van Buchem, the semiretired Dutch couple who had bought a bed-and-breakfast up at Sarrau, ourselves, "the American family." He made special mention of the German couple in attendance today, who had built *la petite maison* on the hill above the little chapel of Notre-Dame-de-l'Aubépine. This modest residence was, he was sure, familiar to us all. Everyone chuckled appreciatively at this tiny jibe, the Germans having built a huge stone house in a grotesque emulation of the old style on top of a hill facing the town and thus changed the skyline irrevocably, and, many thought, for the worse as they now had to look at it. (I have heard the locals call this house "the monstrosity," "that blight," "the German excrescence," "that big stone turd," among other epithets even as they gave grudging acknowledgment of the couple's employment, at top wages and for nearly two years, of so many local masons, plasterers, roofers, carpenters, electricians, and plumbers in its construction.)

Great hunks of the galette were then handed round, a cake yellow with eggs but far less rich than pound cake, with only a bit of powdered sugar and lemon zest sprinkled on top. This seems to be the French equivalent of the Christmas fruitcake, though here people actually are required by custom to consume it in public. I circulated some more, and then we went home, having taken part in one of the few remaining rituals of village life.

That winter, the village of Les Arques was home to just twenty people, more than a third of its houses, their doors and windows shut up behind wooden shutters, empty for the season. The restaurant had closed until March at the end of November, and the museum opened its doors for far fewer hours during the week. The old ones who we used to see in their front gardens or strolling about, they had all retreated into their houses, the only evidence of their habitation the wreaths of smoke curling from the chimney.

Down in the valley at Auricoste, Yvonne's house stood silent and cold, the rain staining its stones gray-black, a few stalks of bolted lettuce the last vestige of green in her garden. As she does every winter when the cold gets too much for her old bones and her single fireplace ceases to warm, Yvonne had moved in with her daughter, Marie-Josée, who lived thirty minutes away near Fumel.

In our little hamlet, at La Mouline, Raymond had long ago put the garden to bed, plowing under the last, frost-withered tomato and cabbage plants. Suzanne's hens had stopped laying, and the caged rabbits had begun to disappear, one by one, into the winter stewpot. I sometimes saw Raymond out and about on his tractor, in the woods. We were doing the same thing, he and I, cutting firewood, though I was limited to those dry chestnut trunks I could fell and piece with a crosscut saw. A blight was attacking the young trees, killing them before they could reach more than four or five inches in diameter, and Raymond had generously allowed me to take what I needed. He attacked the bigger trunks with a chainsaw, leaving behind piles of fresh sawdust bright against the dark ground. I believe I cleaned his hillside of a good amount of deadwood over those long months, in return for which we sat many hours in front of the fireplace which dominated our living room, warm and comforted by the murmur and crackle of the burning logs.

One morning in the middle of February, we awoke to snow, with several inches already on the ground and huge flakes still floating down as the storm slowed. Snow is a rare visitor here, the cause for much consternation and caution on the roads, which were consequently empty. I grabbed a camera and whistled for the dog, setting off up the hill. The village I found perfectly quiet, still as death, not a person about or a dog barking. The roofs were frosted, and the church looked even more severe under its burden of snow. The Virgin wore a white crown, surveying the restaurant across the street as if transfixed by the cheerful ribbon of smoke curling from the center chimney. So well had the village banished all signs of the modern era that, with not a car in sight and the roads unplowed, all it needed was the *clop, clop, clop* of horses' hooves and the crow of a distant rooster to sweep one back a few hundred years.

This is one of the eerier charms of Les Arques, of many places in the more desolate reaches of the Lot, their ability to induce this feeling of abrupt dislocation from the present, and to invoke a past, nearly a thousand years of it, that is at moments almost tangible. I had had other experiences like this before, once walking home from the restaurant quite late on a moonlit night, that shiver up the spine as I passed a ruined house, its empty window frames gaping at me. A flitting shadow passed over my head, and I jumped. I looked up, heard the whisper of air passing through the feathers of the enormous hunting owl I had startled. I had that same sensation now, perhaps emphasized by the concealing cloak of the snow, that I was walking into the past.

An hour later lying on my back in Yvonne's field, I was enjoying the feel of the last flakes melting on my face. Isabelle and I had romped together in the fresh snow, and now she lay contentedly beside me, gnawing a stick. All around us was a landscape of muted sound and color, the rush of the bubbling Divat, which passes through here on its way to the larger Masse River in the valley below, muffled by the gauzy fleece covering every surface. Winter had leached the color from the land, and my eyes saw a dull palette of brown and black and gray broken only by the faint green of cedars and pines on the ridge I had descended. Looking east, I could make out the walls of the village above, the church tower at its center and the single high wall that stands like a sentinel at one end of the graveyard, a crumbling remnant of the other medieval church that once loomed over the valley.

As the warmth leaked from my body into the cold ground, I tried to conjure a more concrete image of what this place must have been like even a century earlier, with nearly a thousand inhabitants, with every house occupied, the roads loud with the clatter of iron-banded wagon wheels on stone, with oxen, horses, and cows steaming in their stalls on every farm, life ruled by the orderly ringing of the church bell. The day was a workday, in this season likely the pruning of walnut and chestnut trees, picking rocks from fields, rebuilding fallen walls, repairing harness and tack by the fire when the cold and wind got too much. Had that life been better in some ways? It had been simpler, definitely, but also shorter, more difficult, dirtier, more uncertain. And therefore somehow more precious or meaningful? What had we lost even as we gained washing machines and running water and indoor plumbing and modern dentistry? Perhaps I was merely mired in *la nostalgie de la boue*, that fine French image that, literally, means nostalgia for the mud, but whose mocking connotations aptly captured the too-easy musings of a cold, wet foreigner hankering for a past he never had in a land not his own before going back to a warm house for a steaming cup of tea and a dry pair of pants hot from the dryer.

I found myself thinking of Patrice Béghain, the former cultural affairs official who had played such a large part in bringing the Zadkine Museum and Les Ateliers des Arques to life, and who had come to love the village so much that he had made it his family's country home, had laid his wife to rest in its cemetery. Patrice is an intellectual in the grand tradition (and there is none grander, some would say, than the French intellectual tradition, where one is almost so-anointed from birth), and sees the country through an intellectual's eyes. Though he has never tilled a field or pruned a grapevine, he recognizes the value of rural ways but sees them from a wider, sometimes more troubling, perspective.

"In fifty years, because that's when the change has been, we have changed our culture," he had told me so many months earlier. "Fifty years ago, France was still largely a rural country. It had large cities and industrial zones to be sure, but it remains in its mind-set and in its thinking a rural country, much more so than England, which had its industrial revolution much earlier. Now France has lost almost completely this aspect except that it remains very profoundly anchored in our psyche. That is to say, the big lie, the difficulty if you will, of the French model is that so many of the French elite are elites who still have somewhere in their heads this nostalgia for this country life. We all have it. Why? Because it was a world of solidarity, of intimacy, of the passing-on of traditional knowledge, of communal customs.

"Of course, you shouldn't overmythologize that world. It was also a society which was very oppressive for the individual. The name of the place where liberty is found for the individual is the city. In the rural world, there are no individuals. You live in a community and are a prisoner of it. It protects you, but at the same time it keeps you prisoner.

"What moves me greatly," he continued, "even though I do not have village roots, is this landscape which man has fashioned, its tracks and its hedgerows ... And all that is so fragile, all the country know-how, the passing on of knowledge ... You can see [that fragility] in the creeping uniformity in the village fête. Over the last ten years these village fêtes have all come to resemble each other more or less, the same band, the same mediocre music, the catered food.

"But it is still a moment where people get together, and that is what is important. It is one of the rare occasions where you still see a rite of community while in the past these communal practices structured the *whole life* of the village whether it was in religion or working the fields together to bring in the wheat or harvest the grapes. All that has disappeared. Even in the rural world, we are living today in a mosaic of individuals. From time to time there are a few moments when the people here can forget the things that have kept them apart sometimes for decades or even centuries and get together. And there is the paradox, for we are living today the end of the rural life that sustained Europe for centuries."

Now thoroughly chilled from a wind blowing raw and damp, I turned my head once again to look up at the village, outwardly so solid and of a piece, so hopeful for its future while innocent, really, of the changes coming down the pike at the hands of a bunch of bureaucrats in Toulouse or Paris or Brussels or Bonn.

Then Isabelle roused me abruptly with a moist whuffling in my ear and a wet stick in my lap, demanding more of the game we had left off earlier. I stood, knowing one thing for sure. As beautiful as this land is, as communally enriching as that life may have been, it must have been bloody cold in those drafty stone houses with only a fireplace, perhaps two, for warmth.

On my way up to the village that morning, I had come across Madame Carrié, a widow bent nearly double with age, she of the famous enormous parsley, sweeping the snow from her stoop. She is old enough to have spent all her life in one of these houses, hauling water and wood, with a privy out back and lamps inside burning oil. Like Suzanne Laval, I'd wager that she could tell you to the month when running water arrived in the village, would show you with pride her washing machine, her oil or gas central heating. Though I'd often heard those of her generation complain that their lives were lived in isolation, their children and grandchildren far away, the old gatherings of a winter eve to crack walnuts or to drink the new wine a thing of the far past, I think they would be hard put to trade the conveniences of today for that life, however fondly remembered. Most if not all have lived their lives partly or wholly off the land, many are of peasant stock, and peasants are above all else practical beings, often among the first to adopt those things that make their day's labor easier.

A few weeks later, on the second Sunday in March, the villagers gathered again at in the village hall, this time an even larger, more disparate crowd. It was the day of municipal elections, when the 152 registered voters in Les Arques would elect a slate of eleven councillors of whom one would be mayor-designate and three his deputies. In theory, a commune could have several slates, and in the larger cities (Paris is a single commune, too) these might represent political parties or alliances of all stripes. (The southwest has traditionally been the heartland of French radicalism, often electing slates of *la gauche cassoulet*, the cassoulet left, because, like this hodge-podge regional dish which can have pork loin, hocks, ham, sausage, and duck, the slate is a coalition of Communists, left-leaning Socialists, Marxists, Trotskyites, and other assorted radicals. The Parisian leftists are called, in contrast, *la gauche caviar*, which should need no translation.)

In practice, the voters of this tiny village had one slate, with Patrick Cantagrel at its head as prospective mayor. Only two of the candidates were new to the office of councillor, Guy Fillion and Jacques. Guy was a recently retired professor of film returning to his birthplace to take up his place in the community. (He is also the son of the last teacher at the village school, and grew up in one of the apartments above what is now the restaurant.)

Since the council and mayor sit for six years, and since they are responsible for everything from keeping the monuments up to the grass down, this is a day of some import. The proof is evident in the sudden influx of those who, though they might live in distant cities, retain the right to vote at the place of their secondary residence. This very sensible practice allows those who have migrated to work in the large cities to retain a real interest in their home-town, a place to which many will return on retirement, which for many in France can be as young as fifty-five years old.

In the weeks before the election, the candidates had set out in groups to visit every house in the commune, introducing them-selves and distributing a single sheet of campaign literature. Jacques had returned from this outing a little shocked by what he had seen. "Some of those houses, Jesus," he'd told me, "it's like the nineteenth century, no central heat, a privy out back, and you have to stand out in the rain to talk to them because they don't open their door to strangers. And this is a commune so small I thought I knew everybody!"

When I had asked Patrick what the big issues were, he hadn't had to think at all. "The roads and the *fauchage*, especially out in the hamlets," he said, "and continuing with the cultural projects already under way here in the village." *Fauchage* is a word that broadly means cutting the high grass along the verges, trimming overhanging branches, and generally keeping the roadside and ditches clean and neat. The importance of this is also signaled in the mention, no less than six times, of these issues in the seven para-graphs that made up the entirety of the campaign literature. "Most of the population lives out in the hamlets," he emphasized, "and they are older, too. They don't want to feel isolated. And they want things to look nice, and not just here, in the village center."

Though *fauchage* may not seem of earth-shattering impor-tance, you only have to drive around the commune in July, a month after the first (and last) trim, to realize its significance. On these one-lane tracks (there are only two roads wide enough to require lines down the middle traversing the entire commune), you need to be able to see some distance ahead in order to find a place to pull off to let an oncoming car pass. When the grass and weeds are four or five feet tall along both sides of one of these nar-row, twisting, gravel roads, your car sometimes feels as if it's sub-merged, coming up aboveground only at intersections. And given the way most people drive around here, you need all the time you can get. The deep ditches lining every road also seem ridiculous overkill until the winter and spring rains come. With little topsoil, and a chalky subsoil over rock beneath that, drainage is a real problem. When it rained heavily for a week, it was not unusual for roads to flood and for some of the deeper valleys to become impassable.

Patrick's platform was neither complicated nor lengthy. "We want for our village to acknowledge its history at the same time opening itself to the opportunities of the present, for Les Arques to bring together its past heritage and its future," he had written. The document veered between these two poles, reaching out to those who wanted continued cultural improvements while reassuring those who lived in the hamlets that their needs would be met as well. It affirmed the proposed council's commitment to the Zadkine Museum and Les Ateliers des Arques, the building and rehabilitation programs that had brought the village into the present with town sewers and streetlights and underground electric and telephone lines. It also pointedly promised better signs and lighting for the hamlets, attention to the all-important roads and the *fauchage*, and lobbying of the regional authorities who had lately begun denying building permits in outlying areas (all the hamlets are outlying by definition) to avoid rural blight and because of the cost of running new electric and water lines.

That was, officially, what was at stake, but under the surface there were several more subtle issues in play. There was the talk about whether the village was going to buy Monsieur Gramon's barn at the currently inflated second-home market price and turn it into a larger, more modern meeting room/recreation hall. The outgoing mayor of nearby Goujounac had done a similar thing, leaving that village of fewer than 250 souls in debt more than one hundred thousand dollars for something many felt was overblown and useless.

There were mutterings, especially among the younger crowd, of the need for a café, a place to gather for a drink when the restaurant was closed, and whether, given their experience with La

Récré, it was really a good idea for the village to try to run such an establishment. This would necessarily involve Jacques, for he held the only—and therefore precious—Class Four bar license in town (which he had bought from the village as part of the restaurant deal), and the authorities were not giving out any more.

There were thorny questions of land use and development, too, especially as there were no longer any houses for sale in the village. Barn conversions were popular (and barn prices accordingly sky-high), but the authorities had begun to issue fewer and fewer building permits, creating tension among those who weren't allowed to sell their property and those who had sold early. The sale of raw land was also an issue here, and one with a particular twist. One sometime resident is a Parisian who owns a large tumbledown stone house outside the village. He seems to be an amiable, if slightly eccentric, sort of environmentalist, and he loves trees. He has been buying up odd bits of empty land here and there around the commune for years. Once acquired, he lets them go, not maintaining the ditches or tracks, leaving the brambles and invading trash trees uncut. This abandoning of the land to nature has created three problems whose long-term implications the commune has only lately beginning to realize. First, if you're a farmer and the field next to yours goes to trees and brush, you will have to spend time and money to keep your land open and clear. Second, such piecemeal ownership can over time create roadblocks to rational development and building, especially given France's twisted succession laws.

Finally, one of the more abstractly valued but quite tangible assets of the Lot in terms of tourism is the particular appearance of its cultivated countryside, that mosaic of farm, field, stone walls, and forest so apparent to the eye from any of the ridges that cut the land here like furrows in a field. Too much forest, and worthless scrub trees at that, would eventually make it a different landscape, one perhaps not so attractive to the northern vacationers, French and European, who have provided the only economic bright spot for the region, and a steadily growing one at that.

224 FROM HERE, YOU CAN'T SEE PARIS

I had thought, with a single list presenting itself, that the election, held Sunday, the eleventh of March, would be a rather dry, pro-forma affair. When I had stopped by in the morning, I found Raymond and Suzanne and a handful of councillors gathered in the main room of the town hall. Perhaps a dozen people were waiting to present themselves, after which they received a ballot and the bright orange envelope into which it would go. These they took behind the curtain in one corner. After they had voted, they slipped the envelopes into the top of the covered glass-sided ballot box, at which point Raymond and Patrick thanked them very formally for doing their civic duty.

"It will be like this all day, people coming and going," Patrick told me at a quiet moment. "But we always have a good turnout, usually eighty to ninety percent," he said. And his prospects, I asked? "*Les urnes parleront!*" he said just a bit melodramatically. The ballot box will speak.

"But it won't be like the Bush election, all that folderol," Raymond added matter-of-factly, having overheard our conversation. "Come back in the afternoon. It will all be decided by then."

At six o'clock, Raymond took a key and ceremoniously unlocked the padlock of the box. People had been arriving steadily for the last half hour, and the room was packed, as it had been at the New Year's glass of friendship. Though many faces were the same, the atmosphere was different. Everyone was quiet, eyes intent on the orange envelopes that the busy hands of the two enumerators were unsealing, then drawing the ballots out of for examination. The ballot inside consisted of the eleven names of the slate, Patrick's at the top followed by his three prospective deputies, then the seven remaining councillors.

Why were they sorting them? I asked Bernard Bousquet, who, dressed in a new pair of jeans, snazzy dark blue wool sweater, and almost clean white sneakers, had taken a seat beside me at the back of the room. "You can choose the whole list," he explained, "or you can cross out those names you don't want or even write in your own. The piles they're making, one is people who chose the whole list, all the rest are *panachées*—mixed ballots either with crossed-out names, or names crossed out and others written in. Very complicated, hunh?" It was the simplest ballot I had seen since voting for class president in high school.

Forty-two voters had chosen the whole list, and so their votes were simply recorded, while another eighty-nine villagers had mixed ballots. Thus, for the next forty-five minutes we sat and listened as, one by one, the mayor read each ballot aloud for the counters to record. There was much more tension than I had expected. At one point Doggie, Gérard Laval's small black cur, snuck in the room and began running about, yapping. Raymond stopped talking, looked up over his reading glasses, and the dog was shown the door. The candidates were being judged, ballot by ballot, in front of us all. Failing to receive an absolute majority and losing would be a severe blow to one's prestige, I imagined, but even the totals themselves were revealing, an instantaneous, and potentially unpleasant, popularity contest with all your peers looking on, people with whom you'd lived and would continue to live for years to come.

As the piles diminished, Jacques, who had started out the afternoon at one end of the counting table, seemed to inch ever closer, until he was directly behind the mayor, from where he could see the ballots and the tally sheet. In the end, the whole slate was elected, Patrick Cantagrel receiving a resounding vote of confidence from ninety percent of his neighbors. Jacques, although elected, received just sixty percent, fewer votes than any other candidate.

"I pass the mayory into good hands," Raymond said when the last ballot had been tallied and the totals announced, smiling broadly at Patrick and shaking his hand. "Our programs can continue, and I was happy to see so many come out to express themselves in this very important way," he finished.

Patrick took the floor, thanking them all for voting, emphasizing, as well, the importance of the turnout. "Next week," he concluded, "to work on the projects we have proposed. I remind you that, in several weeks, we will again gather to welcome the new council to office, to raise a glass to Raymond, who has written such a long page in the history of our village!"

As Jacques and I walked down to the restaurant a few minutes later, I asked him the significance of the vote counts. "You mean," he said, "why I got so few? I know why I got only eighty-seven votes. *J'ai la tête du con!*" Because I can be such a hotheaded jerk! He laughed aloud. "I fish out of season, I pick mushrooms where I shouldn't, and I drive my moto through their hayfields. Christ, there was a time when I told off the whole village council, and you know everyone heard about that!"

At the restaurant, Noëlle joined us as we sat over an espresso in the bar. "Jacques doesn't have the same relationship with people as I do here," she pointed out as we discussed the results. "I am always at the front door, I greet the customers, I ask about their kids, their health. I *have* to know them. He's in the kitchen, busy with the food and so not always as, well, receptive as I am. And when you are part of a small village like this, you pay your respects at every funeral, show your face at every event no matter how small, which is harder for him because he can't always leave the kitchen. But both of us think it is very important to participate, and not only to represent the restaurant, which is after all a significant part of the life of this village."

Looking at the makeup of the council later, it occurred to me that it did, indeed, represent quite a natural balance of the various interests of the village. Bousquet, Lacombe, Tanis, and Vielcazal were all farming families with long histories here. Guy Fillion and Gérard Laval came from the younger, more intellectual, more change-oriented generation. Simone Bladié, Monsieur Laubin, and Yves Astorg fell somewhere between the two, with Jacques a more practical voice of the village's sole business. Patrick, also from an old family, knew them all and was a natural mediator, as well as being a respected speaker who brought with him the broader experience of living, working, and dealing with the outside world from his government job in Cahors, the departmental capital.

Spring brought the promised swearing-in ceremony on Easter Sunday, and with it one of the more unsettling rituals I had yet seen here. In the weeks after the election, the weather, if not wholly changed, had at least moderated. There were days that started in overcast or spitting down rain when I went out in Wellies and a heavy shirt only to come home two hours later shirtless, over-heated, my socks swimming in sweat in the rubber boots.

In the warmth and occasional sun, the landscape slowly shook off its somber winter colors, surrendering to the new growth. The ten-foot-tall Japanese ornamental plum that takes up most of our neighbor Christiane's courtyard burst abruptly into flower, clouds of magenta blossoms almost startling in their profusion. Then the fruit trees bloomed, the peach, apple, and cherry in our front yard thick with clumps of bright pink blossoms. It still rained torren-tially off and on, the creeks swelling, the millponds filling, but at least the rain was warm. In the fields around us, the winter barley and the fallow year grass were green green green, almost intensely so against the still dormant trees.

Many of the hillsides here were replanted in chestnut two gen-erations ago, but the farmers left the occasional cherry to grow among them. These are *merisiers*, cherry trees grown not for their fruit but for their fine wood. They bloom early, standing out like lone cheerleaders in a morose crowd of dark, leafless chestnuts, pleasing to the eye. Forsythia and daffodils bloom at the foot of every wall, while in the vineyards, stark rows of severely pruned vines rise up like gravestones from a carpet of brilliant yellow dan-delions. The rosemary plants, which grow several feet tall and round and so are used commonly as ornamental shrubs, were cov-ered with minute deep purple blooms, drawing bees and wasps and the peculiar, very large moths that drink nectar, and that we had mistaken at first for hummingbirds.

At Hugo's bakery in Cazals the first Sunday in April, Madame Crochard had warned us again that the Parisians were back in droves, and so to reserve our bread in advance. The market, the whole town, in fact, was bursting with life and energy, more people out and about than we had seen in months, and all of them with smiles on their faces and hungry for conversation. The Café de Paris had its tables and chairs out on the sidewalk, and as we sat taking a coffee, two tour buses full of elderly Germans pulled up, disgorging their passengers directly into market, café, and the two restaurants, both of which were packed.

From one week to the next, the number of market stalls had doubled, the pavement nearly impassable with bedding plants and shrubs, rabbits in cages, crates and crates of *primeurs*, the first spring vegetables from farther south. In this region, the bellwether of local produce is white asparagus, which we had been seeing for sale at roadside stands just north on the Dordogne River, kilo bundles of the delicate white shoots at four dollars a pound.

The day before I had met Suzanne out along the road, gathering a trashbag full of dandelion greens for her rabbits. I marveled at that morning's terrific spring thunderstorm, which had arrived without warning, blocking the sun, filling the sky with violence and light, and hurling hail down on the clay tiles until all of us huddled inside (including a very frightened Isabelle) had wondered if we would have any roof left by the time it was over.

It was too bad, Suzanne said sadly, that the hail had arrived when it did. The stones, each about the size of a marble, had knocked so many blossoms off the peach tree that there probably wouldn't be any fruit this summer.

That afternoon I had come upon Raymond out working his vines, replacing old posts with new ones of locust or chestnut, "because they are hard and won't rot for a long time." Piles of grape prunings waited at the end of the rows to be torched. He took a break from his present work, clipping each vine until just a few green shoots remained. Several weeks hence, the tendril would be long enough to tie to the wires, supporting the weight of the bunches of grapes to come. "It's beautiful, that field," he had said, looking across the street at Madame Trégoux's barley. "Such a beautiful green. I remember when you looked at that hillside, farther up, you saw only the rows of vines climbing. Until 1956, when the freeze killed them all. He never replanted." He looked at his own vines. "That hail didn't do my vines much good, either, nor that cold snap." He gave me a weary, farmer's smile. *"C'est comme ça."*

And the birds! Only when they had come home in flocks and were waking us up at seven in the morning did we realize they had been gone all winter. The chaffinches were the noisiest, flitting from the cedars along our drive up to the eaves and back and gossiping all the while. One particularly vociferous couple took to early morning discussions at full volume perched on the ledge of the dormer window of our bedroom, a practice we discouraged by posing one of Lily's plastic Little Mermaid figurines in front of the glass. Triton with his trident kept the couple at bay, at least for a while.

In the winter it had been the owls we had seen most of, the tiny *chouettes* hoo-hooing at night, and the much larger *hiboux*, one of which had flown up from a ditch as I was driving home from Cazals one sullen winter night. With its wings spread, it glanced off the windshield, its wings in fact as wide as the windshield. I only had time to notice that it had had something clutched in its talons before I hit the brakes. I stopped and watched it flap off slowly into the night.

The funny *buses* were back, too, buzzards by the guidebook but looking more ferocious and hawklike than their American cousins. Their wings and bodies are a mottled dark brown, they have the sharp, slightly hooked raptor's beak, and they stand about eighteen inches tall. They really do stand, too, or strut around importantly out in the middle of their fields in broad daylight, sometimes surveying their territory from a roll of hay or a fencepost. My wife called each and every one Mr. Hawkins, and Lily didn't seem to notice that Mr. Hawkins must have been quite busy, flying ahead to be spotted by her no less than four times between Cazals and Les Arques.

The rather macabre ceremony occurred in the late afternoon of a rare sunny, mild spring day that happened to be Easter Sunday, and the day of the new council's swearing-in. The ancient chestnut next to the town hall was covered with what looked like coolie hats composed of pink blossoms, and the villagers were out in good numbers, drawn by the balmy air, the occasion itself, and the promise of tables full of hors d'oeuvres after, courtesy of Jacques and Noëlle.

When I arrived, Raymond was carrying the French flag while Patrick bore a sober bouquet. "We decided," he told the assembled crowd, "as our first official act to lay flowers at the memorial to honor the dead. So we will go down, lay the flowers, say a few words, then come back up to continue with activities a little more festive."

We assembled, processed, and the newly elected mayor, resplendent in tricolor sash (after great argument on whether it should be worn left to right or right to left and assorted asides from the crowd on political leanings) and bearing the bouquet, led us down the main (only) street toward the war memorial. This is a twenty-foot-tall, gray stone obelisk rising up from a bed of roses and bedding flowers inside an iron fence, which is flanked on one side by the repainted Virgin and on the other by a marble remembrance mounted in honor of a group of local citizens who made the pilgrimage to St. Jacques de Compostelle in 1830-something. I had no idea what was supposed to happen next.

Patrick placed the wreath at the base of the monument. He stepped back a few feet and clasped his hands in front of him, his head bowed. I noticed the villagers doing the same. Yves Astorg, in a green sweater and scuffed corduroys, stepped forward, went up to the memorial, and perched a pair of bifocals on the end of his nose. On the east side of the memorial are the words: "To Its Children Who Gave Their Lives for France, From a Grateful Les Arques."

"Soulié, Joseph," he intoned, leaning forward.

"*Mort pour la France*." He died for France, the crowd murmured.

"Verny, Joseph."

· "*Mort pour la France.*"

"Périé, Noël."

"*Mort pour la France.*"

"Jurguet, Marcel."   ·

"*Mort pour la France.*"

On and on it went, Monsieur Astorg pausing only to make his way from one side of the monument to the other, through the astounding twenty-four martyrs of the First World War, the three who had perished in the Second, and the single death, André Rajaud, in the Indochina campaign of 1947 to 1954.

On and on the family names rang out, familiar and well-worn as the old stone all around, most whose blood still ran in the bodies that made up this gathering. In the crowd, I saw Bousquets, Rajauds, Valétys, Souliés, Carriés, Fauchiés, and Coldéfys, all whose fathers, grandfathers, and great-grandfathers we were gathered to honor. At this moment all over France, in many of the thirty thousand small villages just like Les Arques, the same ritual was taking place, the roster of the dead different surely, but the brief moments of remembrance the same. Here, there were no longer any living veterans of the First War, and their widows had passed as well. But the children of that generation there were in some small number, mostly the women. And among these few women, most tiny and wrinkled and leaning on the arm of another, I wondered if perhaps there were those trying to remember a face, a picture of a man in uniform—brother, father, uncle, son—on the fireplace mantel, or a story handed down from happier times.

As I looked around me at the old and young, it struck me that this was the most patriotic moment I had ever witnessed in this country, where people are generally cynical and negative about their government. It struck me very forcefully, too, that this was part of what it meant to live in a village like this, where the names on the monuments and in the graveyards were your own flesh and blood. It contributed to a particular unity of spirit, a strength of purpose built on the sheer continuity of families living, working, procreating, and dying in the same place, perhaps one reason why we were here today, why Les Arques had survived, and was out in force, about to celebrate the taking up of the reins by yet another generation.

After, we all proceeded back up the hill to find trestle tables laid with a quiche and hors d'oeuvre buffet and bottles of wine. Patrick made a short speech before the bottles of bad red and even worse rosé surrendered up their corks. I happened to be sitting next to a winemaker from Bordeaux visiting for the weekend. "I hope their choice of wine," he commented only half jokingly, "is not a reflection of their ability to make decisions."

I was sitting on a wall, scribbling, watching the crowd make short work of the buffet tables. One middle-aged daughter was trying to restrain her elderly mother from inhaling yet another square of salty, cheesy *feuilleté* of something or other. "Maman, they are *soooo salty*! The doctor said—" "Oh, I know," her mother interrupted, waving away her daughter's concern. "Get me another, will you? And one of those quiches?"

Elise Ségol, the widow who lived across from the restaurant, took a seat beside me. Is it always like this? I asked. "*Eh, oui!* Every six years, after they get elected, they offer something to eat, to drink. And it's good, a good tradition to get together, talk, have a glass. You should eat!" she said. "Go take something!" But I just ate lunch, I told her. "So what?" she said. She showed me a beignet covered with powdered sugar. "My weakness. But one more little cake won't hurt you." Later, I caught a glimpse of her in the crowd, laughing with a half dozen of her neighbors, all happily eating and drinking. Our eyes met as she raised another beignet to her lips, and we smiled at each other.

TEN

ONCE MORE, UNTO THE BREACH

When La Récréation opened again for business on Friday, March 2, 2001, Jacques and Noëlle had already been at work for almost a week. After dusting, washing, cleaning, and reassembling all the equipment that had been put away over the winter (when the kitchen had been replastered, rewired, and repainted), Jacques and Fred, his returning sous-chef, had set about their tasks. In a plastic trashcan in the walk-in refrigerator were sixteen gallons of leek and potato soup (using a mere six pounds of butter), and liters of the *fond de boeuf* and *fond d'agneau*, the rich stocks that, much reduced, would become the sauces for the beef and lamb main dishes on the new menu. From coral sauce for the lobster ravioli and honey and lemon sauce for the duckling breast, to the frozen nougat and bavarois desserts, the two had busted their guts to put behind them absolutely everything they could before the first day of service began.

By six o'clock that first night, the kitchen had reached that stage of ultimate chaos that signals prep in full swing. There were half-empty crates and sacks of produce cluttering the floor, and every surface was covered with food in the midst of transformation. At the big center prep table, Jacques reduced perfectly cylindrical tubes of peeled potato into circles against the blade of the mandoline, then overlapped the slices in small rings that, once sautéed, would become the potatoes Anna to accompany each main course. Across from him, Fred was trimming and cleaning Kenyan snow peas for the foie gras salad entrée, while Erik, the new cold entrée and dessert *commis*, whisked vinaigrette and nervously consulted the long list Jacques had dictated that afternoon. At the dish station, Jacques' mother, Jeannette, was running water over a tub filled with salad greens in one sink while snipping the fins from tiny Senegalese red mullet lying on a bed of ice in the other.

I shut my eyes and breathed deeply. My nose was assaulted by the acrid scent of burning butter, beef and duck fat, and hot peanut oil as Jacques cooked off beef *pavés*, duckling breasts, and perch, by the tang of mullet and the heavier lobster-shack smell of the coral sauce. Sweet caramel wafted from the oven, where two *tartes tatins* were baking, mixing with the lighter scent of artichoke hearts simmering in chicken stock. The faint bite of leeks I detected, too, in the soup heating on the back of the stove, and a whiff of honey and thyme, and the perfume of the garnishes, lemon and chervil, in their trays on the long countertop where the plating was done.

I heard the *shweet shweet shweet* of a knife against sharpening steel, the rapid-fire *tock tock tock* of someone chopping vegetables on a wooden board, the double *kerchunk* of the walk-in door opening and banging shut for the hundredth time that evening. In the background Jeannette sang a popular song quietly to herself while from the bar the clink of wineglasses and bright rattle of silverware drifted in. I opened my eyes to find Jacques looking at me intently.

"I'd forgotten what this was like, your kitchen in action."

"They come back, the customers, as if I'd never closed," he remarked, seeming surprised. "We have to be ready. The first week is always the hardest. That's why there's so much *bordel* everywhere." It was true, I thought, that I had never seen his kitchen in such a state of disorder. "I forget things, we run out. The new help—" here he gestured at Erik—"well, you have to come down on them a bit." He rapped a thin-bladed spatula absently against the counter, then turned to flip the potato rings in the big skillet. Behind his back, Fred was rolling his eyes, mouthing the words, "He's *sooooo* slow."

The front of the house had its own routines, with the English waitress, Jeanette, back for the season busy setting tables while the new girl, Alexandra, swept the dining room and then mopped the floor of the bar. In the preceding week, Noëlle's days had been filled with everything from the mundane (ironing napkins, cleaning and restocking the bar, taking apart and cleaning the small dishwasher there, setting up the books) to the gargantuan (ordering, taking delivery of, and stowing thousands and thousands of dollars' worth of wine). "I have the impression," Noëlle had remarked tiredly when I had walked in shortly before lunch, "that we never stopped, that we're just picking up where we left off after a week's holiday."

Reinvigorating the wine list had been her latest project, inviting representatives in, tasting, selecting. She knows much more about wine than Jacques, and he knows quite a lot. "It is true," she explained to me as I went through the new wine list at the bar. "Our cellar was not up to the level of Jacques' cuisine. So we are ordering in a few big Bordeaux and Burgundies, even some bottles of the very best, a Cheval Blanc, a Pétrus, a Margaux, any one of which will sell for two hundred to three hundred dollars. Because expensive wine is not an obstacle for our customers. I want to change the list progressively, little by little because you're still obliged to keep all the other wines, the *vin de Cahors*, the Gaillac and Buzet, even the Nozières at eleven dollars a bottle," she continued. All of these wines are made within two hundred miles, and the most expensive doesn't top fifty dollars. "Our local clients, and some of the tourists, too, expect to find the regional wines even if we're not serving regional food."

Keeping a cellar, especially one with any depth in vintage or region, represents a real risk for small restaurants like La Récré, both because of the enormous capital that remains locked into the bottles until they are sold, and because the government taxes the inventory at year's end. Before, her goal had been to finish out the season in November with her cellar as empty as possible. Now, they would begin to carry over a certain inventory in the hopes that customers would want to order enough of the more expensive bottles to pay the additional expenses incurred.

One change in the kitchen was a photo taped to the wall, a busy, composed plate of sculptured food as one might see in a culinary magazine, evidence of how Jacques and Noëlle had spent a few weeks of their vacation. They had been invited to go to Singapore, to the Ritz Carlton, where Jacques, cooking his own menu, was featured as a guest chef and Noëlle had the chance to join the house staff to see how a five-star, international hotel restaurant was run.

The photo was the Ritz version of a dish Jacques sometimes made here, salmon and scallops over Provençal *escabèche* with a very light, creamy scallop sauce. While his plate was a picture of simplicity, the Ritz's was an architectural monument, squares of pink salmon on top of which were perched the pale scallops, the whole napped with linen-colored sauce. The plate itself was rectangular, the food rimmed with caramel-colored ropes of deep-fried spaghetti (a much-imitated Ducasse touch) and overarching the whole a few wands of deep green chive waving like antennae.

"Not only is that haute cuisine, Jacques," I said, "it's also *la cuisine haute*!" Tall food. Very tall.

He laughed, a bit embarrassed by the big-time treatment given what was essentially a modest dish. "I didn't learn a whole lot about cooking over there," he admitted. "But I had a great time screwing around in the kitchen with the executive chef!"

At the staff dinner (a salad of sliced tomatoes, hard-boiled eggs, and cucumbers in vinaigrette, *entrecôte*, and mashed potatoes), I found myself embroiled in a high-spirited argument over France's new high-tech aircraft carrier, the *Charles de Gaulle*, at present being towed back into port after one of its propellers failed during its first sea trial. "It's cutting edge," Jacques insisted hotly. "I'm not just saying this out of chauvinism, but all of the problems they've been having, it's because it's such a technologically advanced ship. Not like your American aircraft carriers. You're still using carriers built after the last war! The new catapaults, the systems for landing planes, that's where all the problems are coming from, believe me." I smiled knowingly, egging him on with the details of cost over-runs, equipment failures, corruption, and the overreaching egos of the French engineers, all gleaned from a newspaper article I'd scanned by chance the day before.

The talk turned to the season's customers, especially the Parisians who had begun to arrive to open their houses for the Lenten holiday. A certain madame and her husband had reserved for the evening, and Noëlle turned up her nose. "That one, she's a real *pétasse*!" she commented acidly. And that is?, I asked. "A bourgeoise who thinks she's hot stuff, sticking her chest out and expecting favors."

"Oh, my God, some of them!" Jacques added. "They call you at eight o'clock from the car crying into the phone, 'We're on our way down from Paris, and there's nothing in the refrigerator!' So send the husband to the store, lady."

"And that one's in love with Jacques, too," Noëlle added. "I open the door and she rushes right past me to go kiss him. Then she turns to me," Noëlle put on a smarmy smile. "And she says, 'I hope you're not jealous!' "

An hour later, as the evening's diners began to arrive at the door in a steady stream, the kitchen had begun to lurch from crisis to crisis, hitting every little bump in the road on the way back to becoming the well-oiled machine it would have to be by June, when lunch and dinner mean many more meals than the forgiving pace of spring.

At one point Noëlle stuck her head in. "What's the kids' menu tonight, Jacques?"

"There is no kid's menu," he answered shortly from the stove.

"What? You can't do that! I put it on the *carte*," she said, exasperated.

"All right!" he grumbled. "*Filet de boeuf* and some potatoes."

A little later, as the first entrées were going out, Jacques turned from the stove and caught sight of one of the foie gras salads on the counter, ready for pickup. Two thin rounds of foie gras sat atop a mix of multicolored salad greens, cold green beans, artichoke hearts, snowpeas, and mushrooms cut into matchsticks. He probed and fluffed, then shook his head. "It seems a little thin, this salad," he said, going over to Erik's cold station with the plate and adding a bit of radicchio here and another piece of artichoke there. "Can you clean some more greens, Jeannette?" he asked his mother. Then, to Erik, "Like that, a little more substance, okay? And what is this?" He gestured to vegetable peelings and onion skins where Erik had been working on something else earlier. "Clean it up! You don't just do your *mise en place* and leave the crap lying around. You know that!" Erik looked chagrined, and I saw, over his shoulder, Fred nodding.

Later, Jacques would complain that the new *commis* were all alike in some ways, that you had to remind them of the small things, like not cutting orange slices with the knife you just used to mince garlic, as well as the more important elements of the cook's job. Being prepared and working fast were two indispensable qualities in Jacques' kitchen, and he would have no qualms, he often repeated at full volume, about getting rid of someone who couldn't keep up.

No longer so overawed by what I saw going on around me, I found myself able to take a step back, to pay attention to some of the subtler things, the little rituals that marked the restaurant day, the rules, spoken and unspoken, that governed the relationships here. I watched Jacques melting butter into a sauce, how he held the saucepan tilted so slightly, his little finger elegantly extended as the others grasped the handle of the whisk. He leaned down over the pan, lifting the whisk from time to time to see how the liquid slid from the wire, then putting the sauce back on the stove to come up to temperature before adding each pat of butter.

You learn a few things hanging out in the kitchen of a good restaurant, and not only about preparing, cooking, and serving food. You learn a whole new vocabulary of food and of critiquing food, how to deconstruct a taste, how chefs judge each other (severely). About the time the new Michelin Guide comes out in the spring, the phone lines smolder between restaurants with the news of who lost a star and who gained one, complete with vociferous arguments about whether either action was justified, accompanied by intensely detailed critiques of the last meal eaten there. And you hear a lot of gossip from what is, after all, a fairly small, closed world.

One day in the fall, when I had been working in the restaurant kitchen, Jacques prepared beef brochettes for our lunch, dousing them after grilling with the same sauce for that evening's *pavés de boeuf*. We were sitting around the marble table in the bar, Jacques, Fred, Noëlle, English Jeanette, and I, the others having a good chuckle over the latest *histoire croustillante*, or juicy bit. Hélène, the summer apprentice, had left her pastry apprenticeship at a prestigious one-star restaurant in Toulouse after only two weeks because the chef was a screamer who worked his apprentices fourteen hours a day. Worse, and *quel scandale!*, she reported, much to the delight of this crowd, there were cockroaches in the kitchen.

Jacques shook his head. "He discourages his apprentices, this guy. You shouldn't do that. When you take on an apprentice, you can't have them wash dishes and peel lemons all day. They end up thinking, 'This profession sucks!' You can't treat them like salaried employees. They're unpaid, and so you have to teach them something."

"Look at how it glistens, this sauce," Jeanette commented, when the talk of apprenticeships good and bad had been exhausted. "That's a good sauce."

"Of course it's a good sauce," Jacques said, taking his plate and holding it so the the sunlight fell on the meat, its almost chocolate-colored robe of sauce gleaming iridescently from the butter that gave it that particular clinging consistency. He tilted the plate slightly, but the dark puddle barely shifted, only thickening a bit on the downhill edges.

"We say it's glacé, like icing," Noëlle added. "That's a good sauce, how it should be."

"Is there any *fonds de veau* in it?" Jeanette asked, being allergic.

Jacques drew himself up in mock outrage. "Madame, here we don't do that. We use poultry stock with poultry, veal stock with veal, and beef stock with beef! Please."

"Have you ever eaten at Restaurant X?" Noëlle asked, mentioning the name of a well-known place in Cahors. "He uses veal stock for everything, but then, he's not strong on sauces. You go there for the desserts, they're very good."

"*Lui, c'est un pâtissier!*" Him, he's just a pastry chef, Jacques said dismissively.

"Jacques, he's worked a lot in sauces, put in his time with Daguin and Roger Vergé [two of his early culinary mentors], and you can't be a *saucier* for them without learning how to make a sauce like that," Noëlle pointed out simply, Jacques nodding emphatically all the while. "Desserts bore him, not enough room for improvisation."

"Well," he added. "You need a lot of room for *la pâtisserie*, and two people. It's a lot of work. A lot of time. That's why we have simple desserts."

There exists among *chefs de cuisine*, the ones who make the tastes that distinguish restaurants at the same general level of price and cuisine from one another, a rough hierarchy determined in large part by a few gross distinctions having to do with the way they cook. Jacques, who has about as classical an education and background in the kitchen as you find among his generation of chefs, is known for his fish and for his sauces. At any one time, you may find in his kitchen sauces that have as their dominant, base ingredient beef, veal, chicken, duck, lamb, or shellfish. A cut of beef in his restaurant comes with a beef-based sauce, though it may be flavored with *vin de Cahors* or summer truffles, a saddle of lamb with a lamb-based sauce accented perhaps with thyme, perhaps with rosemary. To thine own self be true means for Jacques, quite simply, that he'll stick to his strengths of unusual fish unusually prepared and sauces made, and served, the traditional way.

During service, watching him blissfully crunch a piece of smoked duck breast, lift a few crumbs of walnut to his mouth, touch a spoonful of sauce delicately to his lips, I was also remembering, that first evening of spring at the restaurant, how sensual a chef he was to watch, so uninhibited and natural in his element. He moved very fast, but also gracefully, with an economy of motion, the principal dancer in his own ballet. Outside the kitchen, he could sometimes seem lost, restless, distracted, even a little bored with what was going on around him if it didn't involve action. My wife, who teaches high school English, used to tease him, saying he reminded her of the overactive adolescents she taught, that if attention deficit disorder did indeed exist, he had it for sure. (Not knowing this was considered a negative, Jacques would agree heartily.)

And there were the rituals, like Jacques looking up from his work and yelling, *"Les filles! Du café!"* toward the bar at hour-and-a-half intervals during prep. Five minutes later the tray would arrive and the whole team would gather, at first the only sound the tinkle of spoons stirring sugar into the tiny cups of potent espresso and the *snick-snick* of lighters firing up cigarettes. Each break lasted precisely the time it took to smoke a cigarette and suck down the thick brew, a time during which anyone could ask questions about the day's labor and any problems that had come up could be dealt with.

Another oddity was the almost fetishistic treatment of knives. At one point that afternoon, Erik had turned from his work. "Jacques," he said, holding out a long, thin-bladed knife. Jacques wordlessly took the knife and whisked it against a sharpening steel, then handed it back. I raised my eyebrows, what, he couldn't sharpen a knife? "It's *my* knife," Jacques said, "that's why. And I was kind enough to lend it to him. Everyone has their own knives in the kitchen, and everyone sharpens their own knives. But we lend. Fred, the poor bastard, his knives have been at the grinder's for months now." When one of the chefs acquired a new piece, the others gathered around, oohing and aahing, laying a thumb against the blade.

Knives had stories and personal histories, too. One enormous, heavy-bladed mincing knife of Jacques', its blade pitted and scoured from the acid of ten thousand tomatoes and onions, had accompanied him throughout his entire career. It had sat, forgotten, in the back of a drawer until one day when I had asked Jacques to buy me just such a knife at the kitchen warehouse he frequented. Almost in the middle of service, he pulled out that old *éminceur*, lovingly resharpened it, first against a rough honing stone then a fine steel. He scoured the scale from the old gray blade with steel wool, oiled the wooden handle, all the while telling me of the places, the famous restaurants and those not-so-famous through whose kitchens it had accompanied him on his travels. Finally, he tested it against his thumb, gave it a final wipe, and presented it to me on a clean towel. Following the old superstition, I gave him a silver coin, a franc piece, in order that the blade not cut our friendship.

This warm and spontaneous gesture was completely in keeping with Jacques' character, one of those isolated moments in the chaos of the kitchen when everyone stopped for a second, looking on approvingly, then went back to work warmed by its sentiment. It was genuine as Jacques was genuine, nor was there any resentment that I had been the recipient of his generosity. The two *commis*, the waitresses, the dishwasher, everyone had been wetted by the passing storm cloud of his anger at a job not well done. But their more ordinary experience was of his other side, for he looked out for his employees in the old way, feeding and paying them well while nurturing them professionally if they were receptive. While he cultivated a severe image, his bark was definitely worse than his bite.

The little interchange earlier between Jacques and Erik over the state of his station, with Fred looking on, was completely in keeping with the mix of rigor and gentleness with which Jacques runs his kitchen. There is a good deal of slack, but there are also rules. One sign of this is the way he addresses his help, with the informal *tu*, while they use the formal *vous* with him.

"They always *vouvoyer* me," use the formal address, he explained to me. "I could tell them to use *tu*, but the *vous* creates some distance. And because I am the boss, I need that distance. When we're not in the kitchen, when I'm out riding motos with Fred, I could tell him to *tutoyer* me and he would. And he would know not to abuse it, to be formal when we were back in the kitchen. But with some of the others, they'd be shouting out things during service, and that's not appropriate." He frowned and shook his head. "I don't say anything, they address me formally, and it's better like that. *C'est tout!*" he finished.

After spending so much time with all of them in the kitchen, it had become clear to me that Jacques was ever-so-gradually easing Fred into the role of *chef de partie* (his number two), a rather grand term for such a small kitchen, but with important implications. First, Fred would be responsible for bringing Erik and the other *commis* up to speed, and, more and more often, it would be Fred who directed the pace of the service from the hot seat in front of the big stove. Already, on slow nights in October, Jacques had disappeared upstairs to do paperwork after the start of service and even left Fred in charge one night when he and Noëlle had had to go to a friend's wedding.

Fred's ability to cook was never in question. He knew the ins and outs of prep, knew Jacques' recipes. He had mastered the complex timing of service, could juggle dozens of orders at once and not lose track, make a mistake, or get behind. He shared with Jacques a certain preternatural awareness of the totality of what was going on around him, who was in trouble and with what order and how to jump in to help without being asked. He could do everybody's job, and would, willingly and without complaint, applying the same meticulousness to breaking down the kitchen at midnight as he had earlier to garnishing dessert plates for Erik.

Still, he was only twenty-three years old, his youth showing itself more in a lack of finesse with those he worked with and a certain rough edge to his speech and manner that could and did put others off at times. That evening, when Fred had snapped at a waitress, Jacques had leaned over and said, almost under his breath, "A softer tone with the girls, please, Fred." This wasn't the first time Jacques had given him a piece of advice, and always it was received respectfully, a little nod, a half-ironic *"Oui, chef!"* which is how the underlings in other, more formal kitchens respond to every whim of the big kahuna.

I wasn't sure how conscious Fred was of this ongoing inculcation, until that night in the fall when Jacques had left Fred in charge. Jacques had chosen the evening of Fred's solo flight wisely, for it was the Hunters' Supper, when two thirds of the restaurant would be taken up by Bernard Bousquet and twenty friends, all eating the same thing. Bernard had shot a boar, which Fred had earlier butchered, carefully saving out the liver and heart. The rest of the meat had been cut from the bone and pieced for barbecueing, which he had done earlier over hot coals in the courtyard. The final cooking would be in the oven, just before saucing and serving. It was thus an easier night, since only a dozen or so other diners would be eating off the menu, meaning that much less to keep track of in the kitchen.

Fred, always immaculately groomed, positively shone as he knotted the strings of the white apron he wore over his starched-white tunic, the white-over-white being, in this kitchen, what separated the chef from everyone else, who wore faded blue aprons. He'd even slung Jacques' favorite pair of tongs through his apron strings like a gun in a holster and was swaggering about just a bit at the start of service. He also adopted Jacques' role of teacher with me throughout the evening, explaining that this was a *marcassin*, a young male of two or three years, whose meat would be tender. Boar is always eaten well done, and, with such a strong taste, stands up well to equally assertive flavors, like the black truffle in the sauce that would accompany it. He offered me a crunchy bit of the meat, as well as a piece of the heart and liver.

At a slow point, one of the waitresses had come in. "You guys are gonna love me!" she said. "I just sold a whole table the *cannettes*." "*Bon*," Fred replied, then turned back to his work. Now cooking a whole table of the same order versus six separate items that all have to be done at the same time is a true gift for the chef, the gift of time to do something else. It made the cook's life much easier, and I was a little surprised at the abruptness of his "Good." I must have looked at him a little strangely, because he turned to me and gave the longest speech I'd ever heard him give.

"The other night," Fred said, "Noëlle sold a whole table of venison, which we had a *lot* of, and she came in all bubbling, 'Didn't I do great?' I said, 'Fantastic, Very good!' and after Jacques scolded me. 'Never tell the waitresses, *Excellent!* Or even *très bien*. You can say, *bon* or *ça va*, but never, Great! Afterward, later that night, the next day, they'll make you pay!" The evening went off without a hitch.

Jacques seemed to want Fred to understand much more than just the food, too. That first night of spring, the new waitress, like the new *commis*, was having trouble getting into the rhythm of things. English Jeanette, far more experienced, was left to pick up the pieces and, at one point, lost it completely. She came steaming into the kitchen, red in the face, and spluttered, "Don't send her out there with any more hot plates! You're just dropping her in the shit. She doesn't know where she's going with the food and just stands there in the doorway, looking pathetic!" She ordered the soup for a table of six before loading up once again with hot entrées.

"Wait five minutes on the soup," Jacques calmly told Fred, who was already reaching for the ladle, as soon as Jeanette was out of earshot. "She wants to rush sometimes, gets in a bit of a panic, and you just have to let her calm down. Also, the entrées for that table aren't anywhere near ready." Both Fred and I looked over to Erik's station, where six orders of *croquant de chèvre* were in various states of composition—or decomposition, it was hard to tell. The *croquant de chèvre* entrée is three paper-thin pastry triangles with a delicate round of goat cheese inside, first sautéed to melt the goat then served over a complicated preparation of frisée lettuce, tomato slices, walnuts, melon balls, and artichoke hearts. The edge of the plate is ringed by transparent cucumber slices and the whole then doused with walnut-oil vinaigrette before a final garnishing of chopped green onion. It is beautiful, elegant, and quite time-consuming to make, and Fred immediately went over to help.

Ten minutes later, from the stove Jacques watched Erik saucing a plate of lobster ravioli. "*Non, non!*" He grabbed the plate and dumped the whole thing back into the pot of simmering pasta water. "The sauce is too thick." He whisked a bit of hot water into the saucepot, fished the raviolis out of their water, and started over. "Stop and think," he said shortly. "Ask. No one's ever going to shout at you for asking."

Then Alexandra appeared in the doorway, looking a bit lost, and it was her turn. "Did you put that bottle of Badoit on their bill?" Jacques asked, having heard Noëlle tell her a few minutes earlier to go get one for table ten.

"I'm going to do it," she answered.

"*Non.* Not 'I'm going to do it.' But 'I'm doing it.' Now. Right away or you'll forget," he scolded gently. "And that's twenty francs out the window."

In the thick of service, Jacques appeared suddenly at Erik's side, asking, "So how many entrée orders are you waiting on?" Erik didn't respond. "The last table has been here for half an hour, their entrées just went out, and you should know that because Noëlle called it out as the last order and so did I. So put everything away for the entrées, the salad stuff, the vinaigrette, give your station the once-over with a wet cloth so it's clean, then lay out your *mise en place* for the desserts. Because they're coming, thirty-three of them all at once, and you'd better be ready."

A moment later Alexandra appeared again, chewing a strand of her long brown hair nervously. "Do I," she asked hesitantly, "do I have to order the nougat dessert a little ahead of time?" Everyone stopped what he was doing and looked at her blankly. "I don't know, so they can, um, warm up a little?" By this time the blank looks had turned to outright smiles at her question, which was the equivalent of asking if ice cream should be served a little melted. "Well?" she asked, the first suspicion dawning on her face that her question was perhaps not the smartest. "What's the answer? *Oui, non, où merde*?"

"*Merde*," said Jacques and Fred simultaneously.

"*Merde*," Erik put in, finally in on the joke.

"*Où merde*, pretty much," Noëlle added, as they all began laughing outright. Alexandra stuck out her tongue, then wheeled abruptly and left the room.

As the evening wound down and Erik labored over the last desserts, I took a seat with Jacques and Noëlle at the table in the bar. Over a glass of wine, we talked about their hopes for the new season, what they wanted to change from the last. Noëlle talked about the more varied wine list, the new waitress, how she hoped to let Jeanette run the front of the house more so she could attend to other things, much as Jacques wanted Fred to take on more responsibility in the kitchen.

The menu had changed, too, Jacques pointed out. As always when the restaurant reopens, the fixed-price menu usually starts out with fewer choices than in full season, perhaps three entrées, four main courses, and four desserts. For this first weekend, when not many customers were expected, after the soup you could choose from an entrée of lobster ravioli, small fillets of red snapper over artichoke hearts simmered in bouillon, or foie gras and toasts over a large green salad. The main dishes were Jacques' four standards: duckling breast in honey and lemon sauce, *filet de boeuf* with a summer truffle sauce, saddle of lamb with thyme-scented jus, or grilled salmon fillet over *escabèche*.

And the price had gone up ten francs, too, to one hundred fifty francs for the five-course meal, still less than twenty-five dollars. "With mad cow disease," Jacques told me, "it was hard to make a varied menu at a good price. There was no faux fillet for the *pavés*! Or at least anything at a price I wanted to pay. The lamb is from

New Zealand, the beef from Italy. Italy! Who ever heard of serving Italian beef in a French restaurant? And all very dear. The fish, too, Maine lobsters at eighteen dollars a pound. I don't know if I'll be able to do my lobster ravioli at a price like that."

(There were other surprises, too. Like Jacques' pants. Jeanette had found some cheap, on the Internet, and so he had asked her to buy him a couple of pairs. Chef's pants usually have very small black-and-white checks. These pants look like they were made out of Formula One racing flags, the checks the size of dollar bills, and Jacques was a tad defensive about them.)

Their goals, after the previous year's experiences, had changed slightly, as well. "[Last] summer was different with two cooks and two waitresses," Jacques commented. "It gave me a bit of distance, enough people so things were happening off in the corner or upstairs at night that I didn't necessarily know about, couldn't know about. It gave me a glimpse really for the first time of the other possibility—that this could be a seasonal restaurant. That we could have five cooks and five waitresses and do a hundred or a hundred and twenty covers lunch and dinner for more of June, all of July and August, more of September, then close for four or even five months completely instead of the two, two and a half we do now. This year we're going to close for three months. It's a different thing, really. You're no longer the center of the village then, and I don't know how I feel about that. But the temptation is there."

It is Noëlle who keeps the books, and her perspective was similar, but from a different angle than Jacques'. "This year," she said, "we'll do sixty covers maximum a meal. Last year was something of an experiment for us. We hired more people. We really pushed, both on the number of covers and we pushed our personnel, too. And we didn't make any more money! We worked harder, longer hours, and more money did come in. But the social security charges, they were huge! And Anne, a good waitress, quit in the middle of August, which meant we had to cut back on reservations even though we had an extra chef on the payroll for a whole month after that. In the tax bracket we're in right now, it doesn't make sense to work harder. We'd either have to double our prices or do maybe two hundred covers a meal. It'd be like a factory, which isn't what we do. And then," she finished with a pointed look at her husband, "someone keeps buying motos . . ."

"Hey," Jacques said innocently, "there are chefs who buy wine, have a huge cave. I buy cars, motos. I'm like that."

I went out the door about midnight, my feet and lower back aching from so many hours upright on the hard tile of the kitchen, and my clothes reeking of cooked food. I looked back to see Jeanette framed in the doorway, holding a tray loaded with glasses. As Alexandra looked on, Jeanette pivoted forward and backward from the waist, her free hand pointing out that the tray remained level all the while. She offered the tray to the new waitress, who just smiled and shook her head. She wasn't ready yet, a fact that would be amply proved the next day when she dumped two glasses of Fénélon, an apéritif made with red wine and walnut liqueur, almost on a customer's head, ruining his shirt and shorts. It didn't look like she'd last.

ELEVEN

BERNARD BOUSQUET, ENFANT DE LA CAMPAGNE

This is an excerpt from a journal entry I wrote in November:

I have taken to walking past Bernard Bousquet's farm on Sunday afternoon in the hopes that he'll invite me in. It's usually raining, gloomy, cold as I come up the track past the corncribs, the dilapidated barn with a thousand ducks huddled in clumps in the field in front. I can see lights in the windows of the large, square stone farmhouse, smell the smoke curling from its chimney. He hears his hunting dogs go nuts as I approach with Isabelle, who skitters away from their pen as the beasts, more wild than tame, jump against the wire in a frenzy to get at us. He comes out on the porch, gives me that crooked smile that always makes me think of a ten-year-old boy up to no good. "Tu veux boire un coup?" he asks. Then we sit around the battered table in an eighteenth-century kitchen filled with enormous, boxy 1940s appliances, his four-foot-six mother smiling wordlessly from her chair in the corner next to the stove, getting up once in a while to offer a glass of wine or pastis or muscat or ratafia or mineral water, bottles of which crowd the tabletop. The chairs are taken up by a motley assortment of family and friends of all ages, the younger

men leaning against the walls. The air is close, hot, heavy
with the smell of the farmyard, with tobacco smoke and
wet clothes.

They talk in the shorthand of people who have known
each other forever, whose lives are tied together by blood,
marriage, land, time. They talk of the morning's boar or
stag hunt, how poorly the bottom fields are draining, when
the pig will be killed. There is a moment of silence at the
mention of one of their own, who fell in the woods out
hunting a few weeks ago, hit his head, and died right there,
before anyone could find him, help him. They are respectful,
quiet with me, not knowing, I think, quite what to make of
this stranger in their midst. The afternoon wears on, the
rain continues to fall, and they leave, one by one, until it is
just Bernard and I, and his mother, whose house this is.

He looks like someone out of the French remake of a
dueling banjos movie, Bernard, coming up barely to my
chin, rail-thin, a small head with short gray hair on narrow
shoulders, huge scimitar of a nose under deep-set eyes, thin
lips from the corner of which protrudes the stub of a
Gauloise. The deep lines of his face, etched by hours under
the sun atop his tractor or working the fields, are truly a
map of this very local world, for he is someone who knows.
He knows everything about the local flora and fauna,
knows where the bodies are buried, knows all the gossip,
the local history, too, of yesterday and yesteryear. He
knows who poaches and where, the secret places where
the wild porcinis and girolles grow, who brews the best
old prune and old pear eau-de-vie, which, if you are a
good friend, he might bring you as a present in a Perrier
bottle when he comes to dinner.

The first time I came to visit, we stood around the dim
barn, the air heavy with the reek of animal waste and
something else, a wet-copper smell that made Isabelle trace
frantic arcs across the dirt. Peering around, I saw on a table

a cardboard box filled with cuts of meat, some wrapped in paper and plastic, others just lying there, a whole haunch glistening, covered with a sheen of dried blood. I had caught him just finishing up the butchering of a boar, this room was where he slaughtered, and the unusual smell was blood. He took the haunch from the box, wrapped it in plastic shopping bags, and held it out to me. "For you. From the hunt this morning. It's very good meat, delicious!" I took it gratefully, slung it over my shoulder, and set out on the long walk home in the dark and the rain. I came into the light of the porch at our house to find my sweater black with acrid boar blood, the T-shirt underneath tinged pink.

Bernard's wife, Jeannine, looks so much like him, both in face and build, that they could be sister and brother. She has always had a smile for us, a word, as generous in spirit and deed as he is.

He is very hard to understand, speaking a thick French muddy with a patois that will always be beyond me, which is why I like to see him alone, when I can listen, concentrate. And I've always got loads of questions stored up to ask, which he must find odd, as I get the impression no one's ever asked him to think about his own life before.

He doesn't seem to mind my questions, which can be as mundane as what kind of trees I see planted in such straight rows in the river valleys. They are poplar, he tells me, perhaps a thirty-year investment, and they sour the soil, except that that shows you already to what derisory value once-good land has sunk here. A bitter smile. And the blue plastic mesh bags tied to their trunks a few feet off the ground? They are stuffed with pieces of the hide of deer, whose stench is meant to warn off their kind, to keep them from feeding on the bark.

I asked him once why he had become a farmer. He didn't say anything for five minutes, and then only "Because my father died."

• • •

Bernard and Jeannine, though their farm is a few kilometers away down in the valley, live up in the village right next to the restaurant, where Jeannine can help her elderly mother. Jacques first introduced us one day when Bernard was dropping off a load of produce. I knew he grew a large garden for Jacques, and farmed, but that day he told me he'd soon be out cutting firewood, which he sold on the side. I asked him to bring by a half a cord for our fireplace (woefully underestimating our needs), and he said he'd come by the following Friday around noon, which I promptly forgot.

That day, a rare, sunny, puffy-clouded fall morning in the southwest of France, found me lost in the woods with Isabelle looking for mushrooms. A mile from the house, I happened to look at my watch, realized that Bernard was coming, and raced home to find him there, his tractor pulling a wagon loaded with stinky oak, the small gnarly stuff that is the first thing to take over the fields here when they're abandoned. We chucked logs until the wagon was empty (for some reason they deliver all firewood in one meter lengths here) and talked about mushrooms.

"*Baing, non*," Bernard said. "Probably won't have many cèpes this month." These are the delicious porcini, some as big as a saucer, that prefer the dappled ground under the chestnut groves that cover so many hillsides here. "It's the wrong wind," he continued. "The wind of the north. Too cool, from the north. We need the warmth, from Africa, that's the wind we need." He talked like that, the speech of a shy man, the words dropping in small clumps from his lips as he looked down at his feet, a vague look of surprise on his face, perhaps, that anything had escaped at all. His accent was so thick, his smile so kind as he insisted I pay him whenever.

As I rushed into the house for the money (less than twenty-five dollars for what represented a half-day's labor at least), I thought him a very nice, a very simple man, not realizing until much later that, in this particular meeting, it was I who had come off the rude bumpkin. I hadn't even been there when they arrived (on what was their lunch hour), and I had never thought to offer a glass of wine or a coffee. Worse, I insisted on fussing about with money, which reduced a pleasant half hour's shared labor to a coarse cash transaction. In this situation, people here preferred a few bills slipped into the hand at some future chance meeting, the money almost an afterthought, its significance diminished. You're not going anywhere, and I'm not going anywhere, they seemed to say, so what's the rush? I'm sure Bernard took me, at first, for that most hideous of hybrids, an American who behaved like a Parisian.

Bernard turned out to be a very complicated man, sophisticated in so many ways, wise in so many things, a man hard to pass the time of day with simply because he was always in motion, always working. There is La Brugue, his family's farm, where he grows corn, wheat, barley, and sometimes sunflowers and rapeseed, cuts silage in the spring, hay in summer, and straw in the fall. There, he also raises a few thousand ducks every year for a foie gras company, feeding them up from duckling to maturity, when they go off to another farm for force-feeding. In the spring, he and his wife cultivate white asparagus, a back-breaking job since the rows have to be mounded over and the stalks harvested by hand each morning over a period of weeks to keep the sun from coloring their tips. There is the vast kitchen garden he grows for Jacques each year, and another almost as large beside it to supply their own table.

He also works for others, feeding and culling the herd of fallow deer imported by a retired industrial magnate who owns Faure, an impressively restored estate that lies over the hill behind the restaurant. Absentee landowners hire him to clear brush, thin trees, mow, hay, and till. The Bousquets also own another house in the village that they rent to summer people, and that Jeannine must therefore clean each week. In-between times he cuts firewood.

In his free time, Bernard is a long-standing municipal councillor and, perhaps more significantly, president of L'Association de Chasse des Arques, the Hunters' Club. This is a quasi-official position, for Bernard is responsible to the local game authority, ensuring that the permitted number of the right animal is killed in its season. This is also a prestigious position, for hunting is a rural institution still very much alive. Farmers are traditionally hunters, too, not only because they are born to it, but because the local fauna can be quite destructive of their crops.

(Many times, out riding horseback in the countryside, my wife and I would come across open stretches of pasture and field that appeared to have been erratically and badly plowed. Our instructor, a sturdy country lass named Geneviève, would chuckle. "*Non, non, c'est les sangliers!* It's the wild boar," she would explain. "They come at night and root up the grass looking for tubers and grubs." The extent of their predation can be seen in the fact that there is no season on boar in the Lot, in the knee-high electric fences that surround many cornfields in the fall, and in the number of car accidents they cause on country roads late at night.)

Hunting in Les Arques is still a communal, entirely male activity, and one that has almost nothing in common (except the consumption of alcohol) with the American image of a couple of guys alone in the vast wilderness falling back on their atavistic instincts to get the better of the beast. To begin with, on a land continually inhabited for a thousand years, there is no wilderness, not unless you include abandoned farmland gone to trees. I never did understand fully what or who defined where and when one could hunt, except that it seemed a strictly local affair, determined as much by tradition and history as the presence or absence of the game warden in his green truck. As for its more primeval aspects, here, at least to this outsider, the hunt resembled more large-animal pest control than a duel of man against nature. (I am one of those perfect hypocrites, unable to bring myself to kill Bambi but happy to eat the meat your gun felled.)

The hunt begins at dawn with a convoy of Renault C4s, tiny workman's trucks with a small cab and larger, empty cargo space perfect for a couple of hounds in cages. The convoy splits in two on either side of a ridge, the first group releasing their dogs, who beat the game toward the second, who wait, guns ready, at the base of other side of the ridge generally just off the road where they parked their trucks. The approaching clamor of the dogs signals the arrival of the game, a deer or boar, which appears out of the underbrush and is shot, at which point the first group collars the hounds and all adjourn to the closest farm to skin, gut, and share out the kill over blood-warming glasses of pastis or wine.

It is a brutal, rather dangerous activity, and one in the middle of which you don't want to stray. Several times during hunting season I found myself tearing through the woods, Isabelle at my side, heading for the nearest road at the first, distant cries of baying hounds. Once, walking along the Auricoste road, I heard a splashing in the Divat by Yvonne Trégoux's house, turned and saw a magnificent, panting stag crashing along the streambed. It stopped and looked at me, leaped the road, and disappeared into the woods on the other side just as a line of trucks pulled up. The hunters got out and took up positions facing the stream from which the stag had just emerged, and I continued home with a small smile for the one that got away.

Bernard is Jacques' closest friend in Les Arques (though the reverse may not be true), and it is a relationship built of many parts. Bernard and Jeannine have played a large, if not always obvious, role in the success of the restaurant, especially in the early years. It was not in their patronage, for they are a farm family with neither the time nor the inclination to eat out often. Even today, you are far, far more likely to find them eating at the restaurant out of season, as guests at a winter evening gathering of friends, than ordering off the menu at Sunday lunch.

No, their role was much more fundamental, as good neighbors first who became friends later. It was the Bousquets who agreed to sell the commune's sole bar license, which had been in Jeannine's family for decades from the time they ran a café, to the village, who, in turn, sold it to Jacques as part of the restaurant. This allowed Jacques to make some profit early on when La Récré had drawn the younger crowd from miles around to party at the bar, before it had made a name with its food. The Bousquets seem to have understood right away Jacques and Noëlle's true situation

when La Récré began: that they were barely hanging on, with
every centime going right back into the business and no money for
such niceties as fixing a leaky roof, replacing drafty windows, or
buying a furnace.

"When we lived here those first winters," Jacques told me,
"without heating, only a little wood stove, the both of us sick
upstairs, Bernard came and brought us wood, carried it upstairs,
loaded the stoves in the cold . . . No, if it weren't for them,
Bernard and Jeannine, we wouldn't have stayed."

Jacques grew up hanging out on the streets of Toulouse and has,
through his career, traveled widely around the world while working in the rarefied atmosphere of world-class restaurants. Bernard
has a tenth-grade education and has likely spent his entire life
within fifty miles of Les Arques. Yet each seems to recognize something of himself in the other. To hear him speak, Bernard saw in
Jacques' bluster and restless energy someone who had pluck, who
had the courage to take a risk (perhaps as a younger Bernard, bur-
dened with family responsibilities, never did), and so decided to
help Jacques. The first big meal La Récré prepared was the Hunters'
Supper, only a few months after they opened, the first of many
through the years. Jacques has come to rely on Bernard for advice
about how things work in this part of the world, in the beginning
especially how to get what he needed in what was after all an isolated village wary of outsiders in which he was a virtual stranger.

For instance, Jacques and Noëlle have been thinking about
buying a house, or a piece of land to build a house, in or near the
commune but far enough away from the restaurant so they can go
home and *be* home. Now, in France (and here I speak from experience), the answers to all questions surrounding a given piece of
property—who owns it, what it consists of, if and when it might
be for sale (despite a for sale sign), and for how much, just to
start—seem to exist out there in the ether, ever drifting and evanes-
cent, unknown to the real estate agent especially and sometimes
even to the owner.

In the countryside particularly, where land has been divided and subdivided into smaller and smaller parcels, often over hundreds of years, the owner might turn out to be two, even three, generations of a family of thirteen, each one of whom must agree to all the details. So exploring the purchase of a property begins with what we would consider rather arcane researches involving conversations with the village mayor, neighbors and friends, anybody who might have the least bit of information. Then you approach the owner, who, ever willing to screw the real estate agent out of her fee, not only knows all about you, but has been waiting for your offer. If you're lucky.

One evening in the summer I went up to La Récré after service had ended to find Jacques sitting with Bernard and Jeannine at the table in their front garden. They were having coffee and a bit of old pear and invited me to come sit and talk. After Jeannine had recharged the glasses and cups, passed sugar and spoons, Bernard abruptly said, "Hey, you can't guess who stopped by yesterday."

"No," said Jacques.

"Madame X," Bernard said, leaning back with a conspiratorial smile on his face.

"No shit," Jacques said happily, Madame X being connected with a piece of property near Gindou that currently interested him.

"And the land?" he asked.

"It belongs to her mother. And her mother is still alive. Maybe ninety years old."

"And she can't sell it?" Jacques asked.

"No."

"Is she *liquide*?" Jacques asked, meaning not senile.

"It doesn't matter."

"So, she has to croak, and then she can sell it?"

"There are, alas, seven children."

"Do you have her address? Isn't she the sister who lives in Montauban, or was it Tarbes?"

"Maybe, I have the address somewhere. I'll get it later."

"I mean, I'm only wasting the price of a stamp if a write a letter, right?"

"No." Jeannine said. "I wouldn't do that."

"A farmer I know has the priority," Bernard added. "He's been plowing it for years, and he has the right to buy it first because it is classed agricultural land."

"So he'll sell it to you?"

"Maybe," Bernard answered, a hesitation in his voice. "I *am* a farmer."

"So why shouldn't I just write her, the sister, a letter?"

"You'll drive the price up," Bernard and Jeannine said simultaneously.

"Ah!" Jacques sighed. "Like what happened to Desson up there, Félix offering seventy and Desson ended up paying ninety when he could have had it from the owner for forty-five."

"Exactly."

"So I have to wait." Jacques sat back, deflated.

"Yep."

"Same price?"

"Same price."

"Pour me another drink," Jacques said, holding out his glass.

When you watch the two of them together, it's hard not to see, despite the fifteen or so years that separate them, something of the father proud of the son who turned out well. They banter and bullshit each other, tell bad jokes, drink too much old prune after dinner while gossiping like two old ladies. And Bernard is generous, generous exactly in the way that Jacques is generous, for the simple pleasure of giving.

In the fall especially, Bernard would turn up at La Récré, usually before dinner service when his workday was done. He'd walk in, hands behind his back, then wordlessly put four enormous porcini or a small sack of chanterelles, the dirt still clinging to their stems, on the countertop. And at least once in October, civet (a hearty venison stew dark with red wine and, sometimes, deer blood) would appear on the menu, the venison from which it was made a gift from Bernard. In that wet, wet spring, Bernard had come in more than once with a crate of white asparagus. "Take them, they're not even worth selling," he'd say, and into the soup they went. And it was Bernard, in fact, who had found Fred for Jacques, Fred who would become such an indispensable part of the kitchen.

If Bernard is very proud of Jacques' success, it is a pride reflected in the care he takes with those things he grows for the restaurant kitchen. He and Jacques work very much together in planning what he will grow each year, how much, and when it should mature according to the needs of the kitchen. What began as a small herb garden has taken over most of Bernard's backyard, and each year the turned soil creeps farther down the hill as Jacques grows more adventurous and Bernard more confident.

In the late spring of 2001, the Lot was a world seen most often through a curtain of rain. In the bottomlands along the river Masse, Bernard's fields of early barley were half flooded, the mounded-over rows of spring asparagus a costly disaster. One day at the end of May, in the kitchen, we talked as he dropped off a few more crates of unsaleable asparagus. "Even the ducks aren't happy when it rains so much," he said dourly. "The feathers on their breasts and stomachs burn off with rain like this, because they spend so much time lying on their stomachs and they never dry out. And then the force-feeders aren't happy," he continued, explaining that, when they slaughter the birds, they're left with stubby feathers that they have to burn off, which takes more time. "Everything will be three weeks late this year, Jacques. The *cocos* are in but need sun. The first squash blossoms." Here he scratched his head and frowned. "Maybe the third week of June, if we have some heat."

We went out to look at his garden, now mostly bare soil with rows upon rows of very young plants just beginning to get their growth. "Come back in a month," he told me. "I'll show you around . . . if there's anything to see!"

Bernard's most important contribution to Jacques' cuisine comes not from some heirloom this or fingerling that, but from the lowly zucchini plant, specifically, its flowers. The squash blossoms, simple, elegant, and delicious when stuffed, are not only a summer hallmark of a meal at La Récréation, but also revelatory of Jacques' approach to cooking. They are hard to grow, perish instantly, and require a delicate touch—and lots of prep time—in the kitchen. In order to work with them at all (they are flowers, remember), they have to be fresh. And not market fresh, but field fresh, a matter of hours from plant to plate rather than even a day. Were Bernard unwilling or unable to grow them, Jacques would not be able to serve them, except at an exorbitant supplement.

In the summer, Jacques will offer squash blossoms two ways. The first is as an elegant entrée, which requires the largest blossoms, those slightly smaller than a closed fist and which have the pencil-thin zucchini still attached. I have spent hours watching him prepare them, cupping the open flower in his left hand, careful not to break off the delicate, green-mottled two-inch long zucchini from which the blossom grows. He trims out the pistils, then fills just the cup of the blossom with a few tablespoons of farce made of cooked and puréed scallops and chicken breast bound with cream, folding the petals over gently so they stick to the farce and enclose it completely. He steams them in a shallow bain-marie pan in the oven, with a final reheating in the microwave at serving. The resulting plate appears at table finished off with two tiny crayfish tails and an entire crayfish as garnish, the whole glistening with a generous clinging film of crayfish essence reduced and enriched with cream and butter.

The smaller blossoms will appear as a side dish with every main course, their stuffing of chanterelle or girolle mushrooms along with chicken breast and cream, and sauced with a light reduction of chicken broth and mushroom.

Like many things that appear so simple, the squash blossom on the plate requires a prodigious amount of work and care, both of which start with Bernard, out in his garden at eight o'clock in the morning, which is where I found him at the end of June on the day of our promised meeting.

When I arrived, Bernard and his wife had already been up for hours. They invited me into the kitchen, where a table was laid out with a loaf, a knife, a bottle of red wine, a length of sausage. *"T'as faim? Un coup de rouge? Du café?"* Would I take a glass of red wine, a coffee, some food, he was asking. It was his *casse-croûte*, a midmorning snack for him. "We get up before six to feed all the animals. The ducks, it gets so hot later even by midmorning that they won't eat, and that's my job, to fatten them up before they go to the *gaveur.*"

He drained his glass and stood, and we made our way out back, stopping at a shed to get a wooden crate and a pair of sécateurs. He was wearing espadrilles with no socks, a faded short-sleeve work shirt, and scuffed blue jeans. Once again, I was struck by how lean he was, the only big thing about him his hands, scraped and scabbed, old scars shining pink against tanned skin. He took a crumpled pack of the blue, nonfilter Gauloises from his shirt pocket, offered me one, fired it up and stuck it in the corner of his mouth.

"I've been doing squash blossoms for Jacques for three or four years now," he was saying as we stood at the top of hill on which his house sits, the garden flowing down in rows, some packed with maturing plants, others showing sparse lines of green against the dark soil, until it came up against a cornfield. "At the restaurant, it was a success right off." Jacques had told me that it had taken them more than two years to figure out, after researches with the local farm authority, that it was not so much the variety of plant but how you worked it that guaranteed adequate numbers of the large, vigorous blossoms each day for the approximately four months the plants would produce. "That's true. Only the female flowers are good, the ones with the little courgette." By this time we were in the garden, dozens of short rows of what I took to be young herbs on one side, a narrow path, then four rows of squash plants, each row six feet wide and a hundred feet long. Below us, long rows the entire length of the garden were planted with other veg-etables, tomatoes and some kind of beans the only ones I recog-nized. He went up to the plant at the end of the first row and squatted down beside it. It was perhaps two feet in diameter and the same in height, but I knew that, with summer's heat, it would easily double in size, which is why the rows were so far apart. These were *courgette verte noire Maréchale*, I read from the seed packet stuck to a stake, green-black Marshal zucchini.

"Look at this one here," he said, parting the thick, spiny leaves to show me an immature blossom, its tightly closed petals lime green, with a deep green courgette half the size of my little finger at its base. "That one will be ready tomorrow, open, orange." He spread his fingers wide. "This one is *malfoutue*, too big, mis-shapen." He leaned in and cut it off, throwing it in the dirt at our feet. "Jacques only wants the prettiest flowers, with the smallest courgettes attached. And they're fragile, too, to harvest, because the courgette breaks off. They open in the morning and close in the afternoon. But if you go too early, they're not open yet. Right now they're suffering. They need water badly. The other night, Jeannine and I were out here watering for two hours. Each plant we watered. But we need more rain."

This was another paradox of the Lot, that fall, winter, and spring could be so wet as to drown the fields and flood the roads, the deluge followed by a bone-dry summer, the farmers and wine-makers desperate for a timely shower or two.

Bernard made his way slowly from plant to plant, in his ele-ment, chattering the whole time. "*Mais c'est un travail de patience!*" he exclaimed at one point. Of course, in this work you have to be patient. "The hotter it is, the faster they grow. And if it's cold, the flowers are very tiny." He cupped each perfect flower delicately, snipped the stem, then set it gently in the wooden crate, little zuke facing little zuke.

"It started a week ago. It'll go until October if the weather is warm." He ruffled the leaves of another plant, leaning down, spreading to check for good flowers. He showed me two dwarf flowers, already in full bloom, each the size of a tea rose. "They're too small. But you have to cut them because otherwise they'll grow, the courgettes, and sap the strength of the plant. We want only flowers, more flowers, and no fruit."

He stood and arched his back, stretching. "They'll spread, the plants, and then maybe I'll get two hundred blossoms a day. It goes very quickly. He'll take one hundred fifty a day in full season. And he does *le tri*, too. That's why I do so many, so he can pick and choose." A half hour later, he was finished, the bare soil between rows littered with dozens of blossoms too small, too large, or too malformed, and only twenty-nine in his crate. "That does it. But it's only the beginning."

He showed off the rest of the garden he was growing for Jacques: rows and rows of chives, parsley, basil, thyme, chervil, staked cherry tomato plants heavy with green fruit, and the beans, which turned out to be *cocos*. The big, bushy fast-growing plants were already thick with pods holding the precious *cocos de Paimpol*, white beans named for a town in Brittany.

(These are similar to a navy bean in size, shape, and color but are eaten fresh rather than dried. Jacques simmers them in water or broth and adds bits of prosciutto and chopped parsely for a side dish, or tops them, tossed with cilantro and Cajun spice, with cooked, shelled shrimp or lobster meat for an entrée. Seeing them, I knew the kitchen help would soon be spending every spare minute shelling the twenty-kilo net bags that Bernard would leave once a week on the restaurant doorstep.)

"These are *les potimarrons* for soup in the fall," Bernard was saying, pointing out the creeping vines that produce a reddish-orange, pumpkinlike squash that has a little of the taste of chestnut (hence the *marron* in its name). "*Eleven* rows of parsley, because he always wants more. And the exotic herbs he asked me to plant, horseradish, lemon basil, some others I don't even remember how they're called."

Here he leaned and plucked a leaf, sniffed, made a face. He shook his head, as if doubtful that anything good could come of what was to him a very strange plant. "The horseradish," I remembered Jacques telling me, "Bernard thought I wanted the leaves. I said, no, you let it grow a year or two, then dig up the root, grind it, use it as a condiment. Sometimes I know he thinks what I ask is a bit weird, but he'll always try it."

Did he ever find it hard to grow to order, I asked Bernard, especially when it was for a restaurant? He waved away the question. "I'd do more," he said, "but I have too much other work. It's not that I make so much money. Sure, the flowers are a good little earner. But mostly it's the pleasure of gardening, and for Jacques. They're so nice, Jacques and Noëlle, I can't say no!"

We walked back up to the house, and he stopped at the top of the hill, turning to survey the garden once more. "Soon, it will be time to harvest the barley, then the wheat. After that, then we cut the leftover straw for the ducks' bedding."

I remarked that he seemed to do an awful lot of different things, unlike in America, where monoculture dominates. "You have to say we're far from everything here," he observed. "And our fields are spread out all over the place. Thin topsoil, too. Maybe I'll try some *maras des bois* [a tiny, flavorful strawberry]. The new varieties, they keep on flowering all summer. We used to raise them, everyone did, for the English market. But we stopped fifteen years ago. Too much work and all by hand. That's why we started with the geese. We thought it would be easier!" He laughed dryly. "But that didn't work, the geese, because there were too many Hungarian and Polish imports. So we raise ducks. You have to do a lot of them to make a little. And we cut and sell firewood here and there. *Un peu de tout pour avoir un peu de sous.*" We do a bit of everything to have a bit of money. "But the expenses of living, my God. And after, if you want to retire! My mother, my mother-in-law, too, it's the same. She couldn't afford to keep her house on her own money. The farmer's pension is ridiculous, and the farmer's wife's pension is even smaller. I don't know what's going to happen when there are no more farmers working to support the ones who've retired. They say in twenty years it will be ten farmers working for every seven retired." He shrugged expressively, as only a Frenchman can. "Oh well, for better or worse, we make a living."

Over the year we lived in Les Arques, I did not see Bernard that often, but we did come to have enough rewarding encounters for me to understand why Jacques liked him so much. For one, far from the shy, laconic farmer I had thought him to be at our first meeting, he turned out to be a voluble, often very funny person once he felt comfortable. When I was hanging about at some village event or another in my more or less official capacity, he was always stealing my notebook, making faces across the room, whapping me on the shoulder and urging drink on me, all as if to say, Why so serious? Put that pen away and enjoy!

Once, before we sat down to what promised to be a very good, very large meal at the restaurant, he caught me drinking a large glass of Perrier as all the others partook of a kir. "What—" he paused dramatically, shaking his finger and clucking—"is that you're drinking? You're going to derange your organism. That's no foundation on which to lay a meal like this! You Americans!" The people around us laughed, and I felt my cheeks flushing.

At that moment, I felt quite at home, accepted, happy to be the butt of his humor for the familiarity it suggested. Because he had also, I realized, *tutoyé*-ed me, used the informal *tu*, for the first time. And also, for the first time, I understood what it was to earn that mark of respect. It felt good, very good, one more brief, tantalizing glimpse of a possible future if we were to stay.

TWELVE

THE MICHELIN MUDDLE

One evening when Amy and I were invited up to the restaurant for dinner, we arrived to find Jacques in an unusually agitated and restive mood. It was not some problem in the kitchen, for it was November and La Récré had already shifted to its three-day, off-season schedule, with their ten week winter break just around the corner. As we sat down to a very simple meal of soup, unsauced perch over *escabèche*, salad greens, and crème brûlée, he started ranting about a long and complicated repast he'd eaten at a gastronomic palace outside Toulouse. It had been an invitation from a friend, also a chef, a special occasion for a group of friends.

"That meal," Jacques was saying, "My friend had talked to [the chef] weeks beforehand, so we were all expecting something interesting. Three sauced seafood courses in a row they gave us! Lobster, then crayfish, then scallops—that was too much, too rich, a mistake. And the wine! Six different bottles, two with the dessert. The glasses kept on getting bigger and bigger! You couldn't hardly see the table. And at the end, a *vin de paille*, a beautiful wine which should be served with something light and simple, some fresh fruit, a sorbet, but with that dessert, which Passard in Paris invented and now everyone is trying—carameled confited tomatoes hollowed out and filled with dried fruit ... And then a Grand Marnier soufflé fondant as well! My God, I thought we were going to croak right there."

"You just don't like to eat out in restaurants anymore, Jacques," said Noëlle.

"You're right. It usually pisses me off, it's so fussy," he spat. "Food should be simple, really simple. The freshest, the simplest, none of these frou-frou complicated overdone dishes. It drives me crazy. I just don't find it interesting."

"You're always like this at the end of the season," Noëlle told him. "You can't tolerate anything." She turned to us. "And he's lost weight. You should see him at night! He falls asleep, a big deadweight in the center of the bed. Me, I don't remember any of my dreams, even. It's just too much work, and I'm too tired!"

"And after the meal, talking with the chef, *putain*!" He was off, waving his arms, his voice rising, Noëlle rolling her eyes at him from across the table. "Telling me he's just put two hundred thousand dollars into the restaurant! 'My only interest is in cooking,' he told me. What crap! I like to cook, yes, to prepare what I know well every time, the best it can be. And when I do something new, to do the research, find the right balance, come up with something that works and that I can make work in any situation.

"But then," he finished, "also to go places and try different things, ride my moto through the woods, play with the dog, travel to Singapore or New York or Maine, or wherever."

I couldn't help but feel that something else was going on, and it wasn't long before we found out what it was. He and Noëlle had just gotten back from an unexpected trip, and not a pleasant one. Claude Augé, the brother of his first boss, had died, and they had made the five-hour round trip to pay their respects at his funeral outside Toulouse.

Now, Jacques is not a sentimental person. This is not to say he is emotionally unexpressive or dull or unromantic, as anyone who has spent time in his kitchen (or caught him watching his wife in an unguarded moment) can attest. He takes things as they come, does the best he can, and, together with Noëlle, lives a life of purpose. But Jacques is also someone who has been marked deeply by loss, first of his father in a senseless accident when he was a boy, and then of his stepfather, just after he and Noëlle were married about a year after they opened La Récré. He would never tell you this, not even in his cups, when he is more given to lip-synching Charles de Gaulle delivering his Liberation of Paris speech, complete with grand gestures, than getting sloppy. It is just one of those things you learn walking in someone's footsteps for a year, as I have.

Of the few occasions when I have seen Jacques step outside himself, have seen him release even a bit the unconscious inner brakes, several have come talking about a youth spent in restaurant kitchens. One powerful chord seems to be his apprenticeship, three of the most formative years of his life, in St. Félix-de-Lauragais north of Toulouse at the place where he donned his first *commis* apron, L'Auberge du Poids Public (The Inn at the Public Scales), an establishment as venerable and respected as its chef and owner, Bernard Augé. This seems to have been for him a time of discovery—of his métier, of the wider world, of himself. It is not hard to imagine, therefore, the role that Bernard Augé came to play, in some sense, for a fatherless fourteen-year-old boy out in the world for the first time trying to make sense of his life.

"The best culinary education is apprenticeship," Jacques had told me once, "and one with a good *maître d'apprentissage*, someone who can really teach you, someone who holds your attention, someone who is at the forefront, because cooking has changed so much. I had the good fortune to do my apprenticeship in a very, very good restaurant. At L'Auberge, we were three apprentices, three young guys, and we made a team. With the chef, of course, and it *was* always 'Chef!' 'Oui, chef!' 'Non, chef!' with Bernard. And it's the chef who is always teaching, at least that's the way it should be, so by the time you're in your third year, you can take up any station, from pastry to sauté."

Some measure of how good a master Bernard was, and of how much he was able to transmit his passion to his apprentices, came when Jacques took fifth place in the Best Apprentice in France competition. "It [was] a guarantee of a job," Jacques had explained.

"But I was a boy! At age sixteen, you don't know what's a good job, a good chef [to work for]. And this is where I was really lucky. Bernard Augé sent me to work straightaway in a two-star restaurant under André Daguin, then, inside a year, to Le Moulin à Mougins with Roger Vergé, a three-star and one of the best restaurants in France. That was his goal for me."

While Augé is a regional legend, Vergé is an international star known almost as much these days for having trained a whole generation of top chefs, Ducasse, Chibois, and Maximin among them. Today, in addition to Le Moulin, he runs, along with his friends Bocuse and Lenôtre, the Epcot Center French restaurants for Disney. As if this didn't keep him busy enough, he is also the executive chef at establishments in Los Angeles and Rockefeller Center in New York, a position which draws more on his prestige, overall culinary inspiration, and menu planning abilities than on his daily presence at the restaurant and requires him to sully his apron stove-side more for photo opportunities than as a routine part of evening service.

For Jacques to have landed in Vergé's kitchen a scant year out of apprenticeship, and in 1982, when Vergé was just beginning a creative revolution in French cuisine, was an accomplishement, indeed. It is the equivalent, to reach for an American comparison, of being a first-round NFL draft pick out of high school, and going on to win the Heisman Trophy at the end of the season. Jacques had started out playing with the big boys at quite a tender age, skipping entire years of working his way up, and he was smart enough to realize just to what extent this was due to his mentor.

That November night, as we sat around a big table in the empty restaurant after a simple meal, it was more than an over-the-top restaurant experience that was upsetting him.

"I saw my old boss at the funeral," he said, twirling a half-empty wineglass between his palms. "I hadn't seen him for ten, maybe fifteen years. He was doing a good job of holding it all together when we arrived. And then he saw me, and he just, he threw his arms around me and started bawling like a little boy . . ."

Jacques drew a breath, shook his head as if to clear it, and turned away.

"We talked about the old days, Bernard and I," he continued after a moment, "and of course we talked about La Récré and our life here. That was something, him telling me he was waiting to see me get my first Michelin star, that he knew I could do it, telling me what I should be doing in order to get it. He asked about Noëlle, and I told him she was running the front of the house. 'But you have to hire a maître d'hotel! A sommelier!' he was saying."

"I didn't want to disagree with him, not there, not then, but that's not really what I want to do. Of course I want a Michelin star!" he hastened to add. But, he explained, he and Noëlle weren't sure if they were willing to change their life so completely, which they would have to do, just to make the attempt.

Listening to him talk, I caught the echo of a little disappointment there, and on both sides. As Jacques related more of their conversation, I had the impression that perhaps his old friend and mentor had hoped to see one of *his* boys carry off the colors, that it was on Jacques that those hopes had been pinned. And not only had that not happened, but Jacques didn't even seem to be playing on the right field. For, in France, still today, there is only one field and that belongs to Michelin.

With more than half a million copies sold every year, the Michelin is as close as France gets in sales, if not in spirit, to an Oprah book, a perhaps fitting comparison in a country that considers that the care of the mind and heart begins a bit lower down, in the stomach. To chefs, the Michelin has been the arbiter of arbiters almost since its birth in 1901, a force that, with the appearance of each new edition in April every other year, regularly makes the fortunes of some while sending others into bankruptcy.

(Interestingly, the Michelin was distributed free to gas stations selling its tires until 1922, when, the story goes, André Michelin walked into a garage and found piles of it being used, depending on the version, either to prop up a rickety workbench or the axle of a car. "Men do not respect what costs them nothing," André is supposed to have remarked. Michelin began selling the guide the next year. Equally interesting, it is the only book I have seen Jacques read. In fact, he collects them, an expensive pursuit. When I asked him why, he looked a little surprised. "It's history," he said. "Our history as chefs, the history of cuisine and all the great names and places, how they used to cook. And besides, they're beautiful, especially the old ones, from the twenties and thirties.")

One sign of the importance of the Michelin Guide is that it has spawned its own locutions in French. When the new edition appears, everyone—newspapers, magazines, television shows, and especially chefs—gossips about who got forked or bibbed, who caught a star, was de-starred or re-starred, or double- or triple-macarooned. The blessed are informally known as *les accros*, because they "caught" new forks or stars, and they speak of having "pulled down" or "unhooked" a star, as if grabbing it out of the sky. The disappointed losers always have a chance to go to Paris headquarters to plead, usually in vain, their case ("You're just coasting!" the anonymous officials told one chef who had lost his only star). But then, there is always the next edition, which may be why some see the chase itself as some kind of drug acting only on chefs, and a very expensive drug indeed.

The Michelin restaurant symbology falls into two categories, forks and stars, or macaroons, as they are also called. An entry with one fork is "very comfortable" while five forks is quite simply "luxury." Michelin appears to use the forks to establish the level of the whole experience, bestowing stars for the actual food itself. Thus a restaurant may have three forks and one star, signaling a "very comfortable" dining experience in "a very good restaurant in its category." When printed in red ink, rather than black, the forks promise an even more "pleasant" meal. Just to confuse things further, establishments may also be blessed with a *Bib Gourmand* next to their entry, the little red Michelin Man who signals good food at a moderate price. The most recent edition passed judgment on more than four thousand selected restaurants in France. Of those 4,137 restaurants listed, 407 have one star and

70 two stars, while a slight 22, or less than half a percent, have achieved the three-star level.

The chefs' ambivalence to the power of this gastronomic authority can be summed up by a headline in the special edition of *L'Hôtellerie*, the restaurateurs' trade magazine, celebrating the Michelin's one hundredth birthday: "Michelin, I love you, I love you not!" This is the cry of the chef who, having labored long to win his first star, had it cruelly lifted two years later, instantly evaporating thirty percent of his business and dropping him to the less prestigious "forked" level.

When you listen to chefs talk about what influences the faceless Michelin inspectors who make the ratings, it's sometimes a little like watching CNBC talking heads speculate about what moves the stock market from day to day. A kind of culinary Society of Freemasons, complete with secrecy and symbolism, the whole Michelin enterprise is cloaked in anonymity, its executives very rarely giving even a press interview, and everything about its inspectors, from who they are to how they are trained and what their evaluations are based on, still very much shielded from public view. ("We never reveal our criteria," said Bernard Naegellen, Michelin's director, in a magazine interview, "because that would be the best way to put chefs in a yoke, to muzzle their creative freedom.") This shadowy atmosphere gives rise to rumors, wives' tales old and new, hints of influence-peddling, nepotism, regional favoritism, cronyism, conspiracy theories, political intrigues, even purported biases for or against certain trends and fads in food and decor.

Still, over time restaurateurs have come to know some of the rules, written and unwritten. Once having labored up the ladder to stardom, for example, chefs are expressly forbidden from using the Michelin endorsement in any way (although vaguely celestial motifs have a way of creeping into logos, menus, onto tunics, and into some restaurants' decorative schemes). They are not permitted to signal their achievement on signs or in advertising and are otherwise encouraged to maintain the sanctity of the Michelin mystique by not speculating in public too loudly about the whys and wherefores of their rise—or fall.

These days the press, French and even American, is much given to bruiting about the death of French cuisine, the fall in its standards, the shift in focus away from traditional technique and dishes, the supposed fact that the superb, reasonably priced meal come across by chance at that unremarkable bistro in the middle of nowhere is a thing of the past. My first night back in France after ten years away, in fact, I stumbled on a television documentary about the beleaguered condition of the country's restaurant trade, specifically the crisis at the very top—and at the very bottom, "the McDonaldization of France."

The young reporter seemed to take inordinate joy in his probings of the soft underbelly of the French national identity as he ran down the "facts." He began with the recent bankruptcies of several celebrated Michelin two- and three-star gastronomic palaces whose multimillion-dollar renovations had not motivated sufficient diners to plump down the two hundred to four hundred dollars it cost for a single person to eat there. He followed with the departure of a handful of great chefs for greener pastures in New York, California, and London of all places. A well-known face from the Parisian restaurant world regarded the camera like an old hound dog left behind on hunt day, complaining that the French public just didn't appreciate the finest cuisine any longer and certainly wasn't willing to pay for it. His next words implied that Americans, though essentially bovine in their tastes and equally unappreciative, were at least willing to pay through the nose to be perceived by others as culinary sophisticates.

The middle segments were more depressing still, fleshing out our vision of the country's gastronomic lower depths with a look at the rise of *le fast-food* especially as seen in theme restaurants outside Paris like El Rancho Texas Grill, Hippopotamus, Pizza Hut, and the ubiquitous McDonald's. After views of the exteriors—strip malls surrounded by parking lots that might as well have been outside Peoria as Paris, and lingering beauty shots of plates of unlovely

Mexican slash Cajun slash Italian slash McDonald's food, we were treated to an interview with the CEO of one such chain.

Looking every bit the balding bean counter, the CEO told us that people in the city were abandoning the small shopping districts downtown and hopping onto the ring roads. They were doing their shopping at the mall, and these new restaurant chains were doing nothing more than following their customers. "It's about time," he concluded ominously, "that we recognized that the nature of the French food business has changed. *C'est tout!*"

The climactic moments were given over to footage shot in America, a Hooter's bar in full happy-hour swing, waitresses bursting out of their T-shirts, acres of thigh exposed by the brief, high-cut scarlet shorts that make up their uniform. Could the day be far off, the narrator seemed to suggest, when the French became so distanced from their stomachs that they went to restaurants for something else entirely than the food?

From my experience in villages and towns throughout the southwest and in smaller provincial cities like Rodez, Cahors, Agen, Brive, and Sarlat, both these positions, at least outside the large cities, would appear to be overstated. Simple fact: the nearest McDonald's to us was a half hour away in Cahors, the departmental capital and a city of twenty thousand people. Beyond that, you'd have to drive an hour and a half at least in any direction for your burger and fries. My hometown in Maine, with half the population, has three McDonald's, three Dunkin' Donuts, two Burger Kings, two Pizza Huts, two Subways, one Kentucky Fried Chicken, and one Hardees, all of them within fifteen minutes of my home and most closer. McDonald's in France, despite offering specialties geared to local tastes (Would you like your burger with goat cheese or Roquefort or smoked duck breast, monsieur?), is still struggling to make a profit here, a goal ever receding of late with the huge drop in beef sales after mad cow disease hit the press. If McDonald's is colonizing France, it doesn't appear to be a massive or even impressive invasion, and that the French worry about it to the extent they do shows how hard the fight will be.

As for what's happening in the two- and three-star places, some see it as no more than just another turn of the great wheel, which is just as likely to swing back next year in the favor of the wildly indulgent, wildly expensive, very long and very complex spectacle that is the three-star restaurant experience in France. Others see the failure in individuals: this chef overreached himself, that one didn't keep up with the times, a third thought he could ask anything for his food and people would pay it. And there is more a sense that it is not traditional cuisine that is under threat, but a traditional way of eating. At this time, the cuisine of the moment in Paris is southwestern, and you can't get more traditional than foie gras and duck confit. For us, the whole debate had so little to do with the everyday reality of going out for a meal in the small towns and cities of the southwest that the supposed crisis could have been taking place on Mars.

While Jacques is neither battling it out at the upper reaches of haute cuisine nor contemplating converting to a steak-frites joint, he has arrived at a moment in his career when the pressure is building, from inside himself as well as from others, to make a change. This I know not only from him, but because many have also asked me, as if I had some inside information, if Jacques were soon going to go after a star. (At present, by the way, the Michelin identifies La Récréation as a one-fork kind of place, comfortable, but don't expect too much.)

Our last conversation came almost at our leave-taking, as we drove to the airport one morning in the middle of July, Jacques and I in the van with the bags, Noëlle taking the rest of the entourage in her car. It was a trip he and I had taken many times, to Toulouse to shop the wholesale market. I had sometimes asked to go not because I found the marketing so fascinating, but because it was a time when he did a lot of his thinking, I had discovered, the dark and the empty road unrolling before him conducive to contemplation and, if I were present, revelation.

That morning, the end of an adventure that had been so rewarding for all of us, a summing up seemed to be in the natural order of things. We got to talking about the Michelin business, and all of his ambivalence came pouring out. I reproduce his thoughts here almost verbatim, for they capture his ambition, his dilemma much better than a mere summation could.

I know there are people expecting more. "Oh," they say, "he's going to go after a Michelin star one day, he'll get a star if he puts his back into it." And we could do it. Why not?

It's a whole other way of entertaining people, a Michelin-starred restaurant. You have to create a different context. Take the staff we have now, just the waitresses, say. There's only one who could understand what we were trying to do and help us take it to the next level. It's [having] things like that already in place that would mean we could one day throw ourselves into the attempt to get a star. Because we would need the help.

You can talk to any restaurateur, and he'll tell you, alone, you're nothing! If tomorrow Pélissou [chef/owner of Le Gindreau, the one-star down the road] could only hire ten cooks like my new commis, he'd be serving choucroute garnie and grilled steaks all day. Alone, you're nothing. It's really the people who surround you who make you what you are. Finding them, that's also a talent.

But you have to say, okay, now I'm working only for a star. It's not at all complicated. I could go to my banker tomorrow and say, I want to borrow one hundred fifty thousand dollars. He'd be very happy to lend it to me because he knows that we have a reliable income. Then I go to a kitchen guy and say, here's the money, build me a superb kitchen. I'd go out and buy beautiful plates, crystal, silver, and it can even all be quite simple. I'd ask our new waitress's boyfriend—he's a pastry chef—if he wanted to go for it with us, because now we are weak on desserts.

The sauces I no longer make up in buckets. I have the time to deglaze, to go after the more subtle flavors with this more expensive ingredient or that special sherry, making the sauce right there in the pan the meat was cooked in. But each time I make a plate, I make it a work of art. I no longer do eighty dinners a night, but thirty-five. And for each of those thirty-five, I give it dimension, take it to a different level. I would be working, voilà, only for that star. Maybe it will be the day when I've paid off the restaurant, when I feel liberated from the cost of all the renovations. I think that's a pretty reasonable way to think about it, actually.

But I would be indebted again, and for one hundred fifty thousand dollars! That pressure, that demand, that's what bothers me about the whole thing. I don't want to become a prisoner of my own debt.

And I know, after you decide to do it, it weighs on you, you can no longer permit yourself to do anything else. I know it automatically. And I know it from a friend in Toulouse who got a star and now he's [the chef] become a slave. He has two kids, a house, he is like me, too, in that he likes cars and motos, likes to muck around with mechanical things. And as soon as he decided to chase after a star, that's all finished and he's been obliged to do nothing but pay attention to his restaurant. It took him two years to get a star, he got it, then a year later he was in a total crisis. He told me, "This is not what I want. I want to go out, to travel, to see my wife, see my kids grow up."

I don't want to become a slave. I don't want to be working twenty hours a day year-round in my kitchen. [Am] I going to find myself saying, why am I doing this?

Me, now, if I were to go after a star, I wouldn't hold back an instant. We don't have any kids . . . but still, I wouldn't want to deprive myself of all the other pleasures of our life now. Because I am like that. Now, I work for myself, I want to be proud of what I do, to be happy about what I do, but I don't really care what others think. I just don't give a damn. Truly, and it's because I don't want it to be by the opinions of others that I am guided. There are a lot of chefs in the milieu [of haute cuisine], they judge themselves according to where they are in the ranks of those around them, in the profession, and not by their knowledge of cooking, by their own abilities.

There is something I have noticed more and more, and that's that there are those who say, oh, you've been lucky to have been able to do what you've done with the restaurant. Me, I don't see any luck there. There is a lot of hard work behind it, a lot of reflection, anything but luck. And in any case, I know that in the month of June, there are restaurants around here, restaurants which consider themselves much more prestigious than mine, even if they don't have stars, and they have to be happy doing ten or twelve covers at lunch and twenty-five at night. This when they have room for a hundred at each service. At my restaurant, you can come any day all summer long and it's full. I never have to worry about it.

To me, the star, it's not something you consciously work for. I know I come from the older generation, but I still think it's not something you just go out and get overnight, something that relentless work alone is sufficient to get. It's something that, if it comes, it's part of a career, it's the fulfillment of all the work you've done over ten or fifteen years. Like Pélissou, and I'll give him that.

But I don't even know why I'm talking about this. I don't even want to think about it. And I don't want to think about it because I want to believe we still have other possibilities. I want to finish the work on the restaurant, and then I keep the possibility in mind perhaps of going off to open a place in America, or in Andalusia, or even in Cahors. I need an exit, an out, the possibility in any case of doing something else somewhere else.

I don't really want the star [for me] because that is not the reason I cook. I don't work for the recognition, see? But I know I could do it if the opportunity presents itself.

I do have a great desire sometimes to go do something else, somewhere else. We never thought when we opened La Récré that that would be it, that we would never move on. But now, the resto is almost paid off, we make a good living, and it's hard to walk away from all that. I have my mother here, that's true, and we're very close. But for Noëlle, she has a big family, and her par-

*ents who are after all not so young. For her, too, I think there's no
pressure to change because everything is going so well here in the
Lot. And I would never want it to be that she was following me
somewhere. I want her to want it like I want it, a change. She is
my wife, and everything we have accomplished, we have done
together.*

When he's not in the kitchen, you can sometimes find Jacques
on his Suzuki all-terrain vehicle, ripping around the farm roads
and forest tracks of the Lot. Or you might, in the afternoon, come
upon him as I did one day, seated on an overturned bucket in the
backyard cooing softly to his truffle oak, encouraging it to grow
fast and bring the truffles. While truffles do prefer to grow in the
soil underneath this particular kind of oak, as Jacques' tree is only
six inches tall and not in a traditional, truffle-loving ground, he's
got a long, perhaps fruitless wait ahead. Yet that doesn't stop him
from crooning to his little tree when he has the chance.

"Cooking isn't the only thing I love," he told me once. "I love
traveling, riding motorcycles, fishing, entertaining friends. I love
life! The great ones of cuisine love only *la cuisine*. They live only
to work. And after work, it's writing cookbooks, inventing
recipes, promotions with food producers, wine shows. I don't
know if that's true for all of them, but for the ones I know, it's *la
cuisine, la cuisine, toujours la cuisine*. Today, with this restaurant, I
sometimes realize I should perhaps be more responsible, but . . ."
The words trailed off. He shrugged, a wicked grin lighting up his
face, and I could suddenly imagine the adolescent troublemaker
his mother has told me he was. "What would be the point?" he finished at last. "*La cuisine* is how I earn a living. I love it, absolutely.
But after, I live . . ."

Much later, when I had long since returned home to America
and was putting together this book, I came across the copy of
L'Hôtellerie dedicated to the Michelin Guide among my papers.
Jacques had given me the magazine in the kitchen one day, throwing it across the counter and saying, "Here, this might interest

you." I reread the interview with Bernard Naegellen, Michelin's director, in which the reporter asks what advice Naegellen might give to an aspiring chef hoping to pull down a star one day. "Be yourself," Naegellen had said. "Make what pleases your clients. Above all, never cook with a guidebook in mind." In the perhaps twenty-five hundred words of the whole text, Jacques had picked out these sixteen alone, highlighting them in bright pink. They could have fallen from his own lips.

THIRTEEN

FROM HERE, YOU
CAN'T SEE PARIS

On the map, Les Arques appears to be almost halfway between Bordeaux and Toulouse, two of France's larger cities. The village, it is true, rates the most miniscule of dots befitting its small population, but around it there seems to be such a thicket of other, larger towns that the area looks well peopled. Once you begin to climb the *causse*, that first major limestone ridge just north of Cahors, however, you realize that the map has deceived you. The landscape begins to unfold, ridge and valley and village, then up another ridge and down into another valley where lies another tiny village, and so on, endlessly it seems in all directions.

When we first arrived, it all looked the same, every village just another dusty, secluded, shuttered-up collection of weathered stone houses and dry-set stone walls by a river or on a hill, the surrounding countryside of field and forest dripping with a kind of beauty and timelessness that we associate more easily with a Millet painting than with a July afternoon in the twenty-first century. We always got lost, always, one reason being that French signs trumpet not the number of the route or road as we are used to, but point (literally, as one end makes a V like an arrow) instead to the next towns and villages, the importance of each signaled by the relative size of the letters in its name, its distance deduced from its position on the signpost. This is fine once you get to know the larger towns and small cities of the region, but in the beginning you find yourself scratching your head a lot and saying, so where the hell is Castelfranc or Fumel or Gourdon, anyway?

I had also never lived before in a place in which everything was so exquisitely and extravagantly monikered. We lived in the region traditionally called Le Quercy, in the part of that region known as La Bouriane, in the canton of Cazals, in the commune of Les Arques, in the hamlet of La Mouline, and in the house called La Forge. At first, we thought, how charming, our house has a name! until we realized that *every* house has a name. Sometimes, it seemed, every pile of stone, remnant of wall or chimney, even every place where a house was rumored to have been if only a whisper of a ruin were left, all these had names, too.

We would often be driving along in the middle of nowhere and see a gravel lane forking off into the fields marked only with a stubby rectangle of white-painted metal perhaps six inches wide by two feet long with a single word on it in dark blue and an arrow. It would say "Lestour" or "La Borie Bas" or "Auricoste." These, we learned eventually, were not villages or even hamlets, but the names of a particular place, a stretch of land, a specific farm or house its anchor, often former family properties, some of which hadn't existed as such for hundred of years. Yet the names have endured, dragging along with them a whole tangle of history, much of which the people who live here still know, if you bother to ask. Over time, of course, the names become very important, as much a part of the landscape as the pieces of ground they speak of and sometimes the only lingering reminder of what once was.

La Mouline, our hamlet of six houses, for example, means the place of the mill, for there were five of them stretched out along the course of the Divat and the Masse. These mills were the backbone, along with grapes, of nearly all economic life in this valley for almost five hundred years. They ground grain, felted wool, pressed oil from walnuts, sawed lumber, and, closer to our time, pumped irrigation water and made electricity.

The last mill in the line, also the largest, is just across from Raymond's house next door, having been a smelting mill until the end of the nineteenth century. The water wheel ran a bellows to superheat the charcoal from the local trees, which in turn melted out the iron from the rough chunks of ore mined in the surrounding hills. (Riding through the woods in Goujounac, we had come across piles of iron ore abandoned where they lay a hundred years earlier.) Some of that iron had been forged in the barn across the drive from our house by a long-dead blacksmith, *un forgeron*, into the rough hinges and hasps we saw on the village houses, ours included, hence the name of this place, our place, La Forge.

On one of our last nights in Les Arques, we came home to find Yvonne Trégoux and her daughter, Marie-Josée, sitting on our terrace, waiting to say goodbye. The terrace looks over the mayor's garden, and, on the left, the end of the Auricoste road, and at this hour of soft sun and shadows, it was quite stunning, the greens of the walnut trees and grape vines and Marie-Josée's barley fields, the ruddy earth where Raymond had turned over a patch of soil. Soon, Marie-Josée was telling stories, about her father, who had used the former smelting mill to saw logs, about another man who had made clay roof tiles in the little fallen-down barn we could just see at the corner of the barleyfield.

"And this, where we are sitting, a *sabotier* worked here," she continued. "He made sabots, you know, the wooden shoes, out of big blocks of walnut. People were very proud of their shoes when they were new, those who had them. I remember my grandmother telling me once how she got a new pair. She didn't wear them in the fields. She didn't wear any shoes! Can you imagine. She showed me how, when the fields had been harvested and all was stubble underfoot, the way you walked was sliding your feet along, to push down the stalks so they didn't hurt your feet. She would carry her new shoes and then, when she got to the edge of the village, she would put them back on!"

I looked down at her own feet, clad in broken-down carpet slippers, at her face, still smiling a gold-toothed smile, her white hair like clouds backlit from the sun. What a character, people always said of Marie-Josée, what a character! I remembered almost the first time I met her, it had been in the fall, Marie-Josée out raking and burning leaves in front of her mother's house clad in pants down below and a black brassière up top. And later, how she tried to coach me in mushroom hunting, how and where to find the various kinds, all without letting slip a single place within miles where I might look.

That they had stopped by that evening, and especially lingered until our arrival, was unusual, a benison, their stories a last, unexpected gift on departure. Others had given us pots of their own foie gras, the Lavals a riotous supper, Jacques and Noëlle a last dinner at La Récréation. We had by then, it seemed, been saying goodbyes all over the canton for weeks, to our English and French friends and acquaintances, to the land, too.

Each one of those last days was tinged with a moment or two of bittersweet, as we did things, went places, saw people we knew were soon to disappear from our life, at least in this particular way. Yes, the path through the chestnut woods over the hill behind our house would always be there, but I would no longer know it intimately, in the fall rain, or on a blowy, winter day with the barren trees dark against a brilliant sky, wouldn't see the incremental changes as the wildflowers come and go, that moment in June when the sweet smell of hay seems to perfume the entire little valley as it does when the farmers all race to cut their fields before the July rains.

In the van on the way to the airport that very last morning, Jacques had, in an odd turnabout that I didn't even realize at the time, interviewed me about our year. "You leave happy, yes?" he asked. "You did what you came to do? Was a year enough?"

"For an idea like I had," I told him, "it's always at the end that you realize, here's the moment when I know everyone, everyone knows me, when we're just beginning to have a deeper relationship with the people we've met. I could have gone further with them, but it's always like that. I will always want more."

"It's true that, once you've left," he said, "it will be a clean break. When you arrive in Boston, at home in Maine, tomorrow, that will be a clean break. And that will be the start of another kind of work, of memory, contemplation, and you will have to find yourself again, in your life, in your work."

He is a wise fellow, that Jacques. In struggling to make sense of that year, what my wife aptly called "a time out of time," it took me some months before I could even bring myself to look at the many photographs we had taken. We were both depressed, she and I, and taken with sudden rages. I found myself cursing aloud in the supermarket one day at the shitty cheeses on offer, staring at the people around me, thinking, How can we live like this? My wife railed nightly against the consumerism that saturates every aspect of our lives to the point where we don't even realize its presence any longer.

In comparison, our life in the backwoods of France was achingly simple, a culture without all the distracting trappings of consumption that struck us so unpleasantly on our return. Our experiences seemed purer, somehow, distilled, perhaps because our immersion had been so total, our acceptance by others so quick to come and rewarding in its depth.

The Lot is unique as a landscape, Les Arques a whole world writ small. Here, I saw lightning go sideways in a summer storm, picked wild porcini, hunted truffles with dogs, harvested walnuts knee-deep in mud, drank plum moonshine, watched as a hailstorm ruined a summer's fruit in ten minutes. Here, my daughter had learned to speak another language, held a newborn goat, nursed a chaffinch, lost her first tooth.

My wife is a very private person (hence her absence in large part from this book), and the absolute last way she wanted to be identified was as the writer's wife. Without a defined role like mine, her experiences were quite different, more interior, as she lost herself in her own writing and long walks with the dog. On these rambles she met far more of the backcountry widows and farmwives than I ever did and came to know the village and its environs in a much different way, too. As the year wore on, I continued to meet people who would say, yes, I know you, you're the husband of the woman who walks with the dog. That seems to have become her identity to many there, the woman who walks with the dog, and a high compliment to her as she prefers the company of most dogs to that of most people.

Sometimes, it seemed to me that I spent my days asking questions, trying to fit together the pieces of the puzzle whose edges were ever-expanding. Then pure experience would intervene—an hour with Raymond tending his vines and his trees, a frantic evening in the kitchen up at La Récré—and notebook and pen would be forgotten.

"There is a moment," Jacques had observed on our last ride, "when you have to let go, to stop asking why, because it's just like that. Even we can't explain why things are the way they are too well sometimes, except to say, that's the way it is. It's in our customs, our culture. Some questions have no answers."

To which I would add that my direct experience always led to a far deeper level of understanding than mere observation ever could have. If there is one thing that surprised me about our time in and around Les Arques, it was that a life which at first seemed so simple could be so rich in such experiences and that the people were so eager to include us in them.

There is in French literature a term called "the American image," which connotes vastness, wildness, the land of the savage in all his unspoiled nobility. In contrast we have the notion of the village in the rural landscape, a closed-up place of isolation and, perhaps, desolation, old people, entropy. As I conjure in my mind's eye those few square kilometers of ragged hills and field and forest that together make up Les Arques, I conclude that, even in a year, I learned very little. Every path has a story, even if only whose oxen made it on their way to what field to work; every wall has a history, even if only of who built it and when and why.

Though much has been forgotten, the people of Les Arques, and of the Lot, especially over the last generations, have begun to realize the value of their heritage and a heritage largely preserved in its unique landscape.

The renaissance of the village today is due almost entirely to what happened there yesterday and the efforts of a few over the years to honor that. This they did by not letting the church fall down or the schoolhouse go to ruin. The farmers continued to farm, keeping the fields from going to brush and rebuilding the walls, and the authorities have been smart enough to see that agriculture in the service of tourism is not necessarily a ridiculous idea.

It is very easy to romanticize rural living in France in writing, but when that image becomes a reality, the result is even sadder, even uglier. There are places in the Dordogne just to the north, and in Provence especially, villages restored to within an inch of their lives, every courtyard planted, every roof tile in place, that conform to our ideal image of what a French village should be. But there are no French people living there any longer, and the villages are largely deserted outside of the summer months. They have ceased to function as a village and instead have become vacation colonies.

Les Arques could have become a lovely museum with a showcase restaurant and church and populated largely by foreigners and summer residents, who can always outbid the locals when a house comes up for sale. (One frequent complaint we heard from young couples from the area was that not only could they not afford to buy, but landlords could make so much money renting at premium rates for the two months of summer that year-round rentals were nonexistent.) Though perhaps a quarter of the houses have indeed passed into such hands and are now only sporadically occupied, the balance has shifted no further because there is no longer anything for sale in or around the village. This is a healthy sign, for it means that the people who live there want to live there, that they are hopeful about the future.

As for the farmers, the Bernard Bousquets and Luc Bories of

the region, they haven't given up either, despite the European Community's attempts to put them out of business in the name of efficiency and the imposition of a single agricultural policy. Luc is going organic and is prospering, albeit modestly, with his ducks. Bernard has begun to sell his squash blossoms to other restaurants and is contemplating growing strawberries next year, too, as well as adding to what he grows for Jacques. Indeed, the French, after the onslaught of mad cow and hoof-and-mouth disease, seem to be ever more appreciative of small producers just like these two, and willing to pay the price that comes with knowing the one who grows your food by name.

While I am hopeful for Les Arques, I fear for rural France as a whole. France is a small country with nearly sixty million people, and it is surrounded by even smaller countries with relatively larger populations. Regions like the Lot, with reasonable real estate, lots of sun in the summer, and exquisite natural beauty, are under pressure, the inhabitants of their villages greatly tempted when they can no longer make a living from the land.

And then there is the idea of the village itself, a fragile concept that many of our notions of modern life work against. The world today is about speed, technology, change, material gain, a place where tradition is often seen as an impediment to progress, where the individual seems to trump the collective every time. The life of a village, on the other hand, is about community, about respect for the past, about the greater good for all, about continuity. I am not sure the two visions are compatible.

FOURTEEN

HOW TO EAT WELL IN A
RESTAURANT IN FRANCE

Les meilleurs cuisiniers sont souvent cabochards.
The best chefs are often pigheaded.

—FRENCH SAYING

By the time I left France in July 2001, I had spent thirteen months
in and around Les Arques and its restaurant, La Récréation.
Sometimes I was in the kitchen during the high tide of service,
perched in the corner taking notes, or during the calmer, quieter
morning prep, asking questions, absorbing everything from knife
technique to ravioli secrets as I watched the alchemy of Jacques'
cuisine. I put in time at the pot sink and racking dirty dishes, time
over a prep table cleaning scallops or red mullet or hacking up
pumpkins for soup. I had hung out in the front of the house, peering over Noëlle's shoulder or pestering Jeanette and Sandrine and
Alex, the waitresses. I had eaten dozens of meals with the staff,
sharing in their gossip, their stories good and bad, talked with
Jacques till midnight at the marble table in the bar, doing the post-

mortem on the evening as the last customers lingered over coffee and old prune. And I had consumed Jacques' cooking at his table in all seasons, both as a customer and a friend invited up for dinner. Together, with our wives, we had eaten out in starred gastronomic palaces and paper placemat auberges, in bistros in Toulouse and tapas bars in Spain, in French fast-food joints and at one of the best restaurants in New England (Fore Street Grill, in Portland, Maine, for the curious).

I learned a few things. I will never know as much about food and cooking as Jacques Ratier. I will never wield a knife with such careless ease, sling a sauté pan with such élan, or be able to compose a restaurant menu, from soup to café, in my head and then make it a reality, night after night, 140 times, most of them nigh-on perfect sometimes for three months running with less than a day off a week.

As for the front of the house, which is where the restaurant begins and ends to most of us, here I found a kind of nightly magic act taking place, order out of chaos, with Noëlle the chief conjurer. She fills the room with warmth, her words like balm on rough waters. Her vast knowledge of her customers and their wants, the invisible but enormous concentration she brings to her work in the name of your good meal . . . I bow down my head in obeisance. She also knows a good deal about wine.

In none of these things, sadly, will I ever be the equal of either one of them.

My privileged position as neutral observer did, however, give me a unique perspective on how the restaurant functioned each and every noon and night I was there, when it went badly and why, and when it was a dream and why. I am a lifelong Francophile, of reasonable means, and do not consider myself unsophisticated when it comes to dining out. Yet standing in that neutral ground between kitchen and dining room, I was constantly learning new things and stumbling on truths timeless only to those who knew them: nothing is obvious to the uninformed. And the uninformed here, many foreigners in France, sometimes didn't enjoy the best meal they could out of cultural misunderstandings and simple ignorance of what was expected of them as customers.

Here, very briefly, is how to get a good meal in France. The point of departure is acknowledging that yes, you are in France. You are here, presumably, because France and the French as a foreign culture interest you, because you want to experience in full one of the pillars of that culture, their cuisine, and because you are hungry. We come from a land where, in theory, the customer is always right, fast food is most food, and we are used to eating anytime, anything, anywhere, and anyway we like it.

In France, the customer brings to the table a knowledge of cuisine honed from earliest days. This education starts with his first restaurant experience, the *cantine*, or school cafeteria, usually at age four. At my daughter's school, the menu was posted outside for the parents to read, and it was invariably a revealing document. The meal is served in courses and contains a dazzling variety of meat, fish, and vegetables prepared in all sorts of ways. Thus are the French inculcated, from an early age, in the culture of eating, but with the expectations not only of what they can expect in a restaurant but what the restaurant expects of them.

Fast food, outside the big cities or the rare McDo, as they call McDonald's in France, is any meal that can be eaten in under an hour and a half. And not only do the French celebrate each food in its season, but they also tend not to eat in the car, or while walking on the street, or to violate the general principle that lunch is at lunchtime and dinner at dinnertime, and snacks only for kids after school. That last, coupled with the exercise of moderation in how much they eat and the higher quality of their food, may explain why obesity is not the national problem it is in the United States.

The Michelin Red Guide, the Gault-Millau, the Routard, the Blue Guide, Lonely Planet—all these resource books exist for a reason. Use them to find the kind of restaurant you want to eat in wherever you happen to be and then *make a reservation*. When the chef knows how many are coming to dinner, he can adequately prepare—and shop. In many small towns and villages, there is no store to supply what runs out, providing the chef could get it there anyway.

At La Récré, Jacques and Noëlle know exactly how many people they can feed and plan accordingly. When they begin approaching max numbers for the week ahead, Noëlle begins to check with Jacques each time the phone rings. Jacques may stop and think, he may go to walk-in just to confirm how much there is of what before saying yes. If it's at the last minute, he may say, "Yes, but they'll have to take the salmon or the *pavé de boeuf*." Sometimes, he just tells Noëlle to stop taking reservations, period, because it's been a long week and everyone is tired.

A reservation in this country is a contract: the restaurant is obligated to serve you if you show up anywhere near the time you booked. You may also eat better, discovering a lovely pile of local cèpes or the squash blossom stuffed with scallop mousse on your plate to accompany your main dish. Last minute walk-ins, provided they're seated at all, will have to make do with sautéed zucchini or an extra stuffed tomato since both the mushrooms and squash blossoms are prepped each morning in the numbers required.

And when you call to reserve, don't call during service, when every available hand is cooking or serving the full house. Call after ten but before noon and after five but before seven or eight. That way, you're sure to avoid screw-ups. This is also the time to express special requests: an extra course for a special occasion, a vegetarian meal, a plate for a child allergic to chicken.

Don't show up a half an hour early and expect to sit at the bar and be served as you would in the States. Many smaller restaurants don't have a bar in any case because it requires a different, and expensive, liquor license. In general, French restaurants open at twelve or twelve-thirty for lunch and seven-thirty to eight-thirty for dinner. The half hour before they open is almost always when the staff has their meal. They've been doing the *mise en place* since eight that morning and will work through to three or three-thirty before the whole thing starts all over again for dinner at six. So don't steal their half-hour time to eat, gossip, have a cup of coffee and a smoke. You'll eat better if they're happier. Call ahead if you are late, even from the road. Otherwise, you may be asked to order from a restricted menu.

If you don't have a reservation, get there as early as possible and offer to eat the *menu du jour* or even *plat du jour*. Smile a lot and chat up the mâitre d' or hostess, who is usually the chef's wife. Do not pull out your wallet and attempt to bribe them into serving you. France is not Russia, and such behavior is considered offensive.

As well, when you make a reservation, you are in general buying your table for the evening. It has not been double-booked; there will be no ten o'clock turn. Nor will a waitress be hovering anxiously to clear every course as soon as possible in the hopes of hurrying you out the door so she can seat the next group they've booked who are already on their second cocktail at the bar. A leisurely five-course meal is easily a two-and-a-half-hour adventure in France, and there is no way at this pace to turn tables unless the chef doesn't mind his staff still cranking out desserts at one in the morning and breaking down the kitchen at two A.M., in which case he'd soon be working alone.

Reservations are actually becoming more, rather than less, important in France since the introduction of "the law of thirty-five" in January 2001. This law, which was intended to create more jobs by reducing the hours in the average workweek from thirty-nine to thirty-five (with employees still getting paid for thirty-nine!), allows restaurants a little leeway, but still permits them to work their employees only forty-three hours a week before the owners begin incurring massive supplementary social security tax bills. And, beginning in 2002, they must offer their employees two successive days off every week. This has put many restaurants, especially in the middle and upper echelons (where they cease to be family operations), in a bind because the typical sous-chef puts in something like seventy hours a week in high season and waitresses not that much less.

The net effect, rather than creating jobs, has been that more restaurants, especially the fancier ones, are closed more often during the week and for longer periods. In the less fancy places, the chef and his or her family are putting in more and more hours since they cannot afford to take on additional help. Or the restaurant simply closes for at least three months a year completely, which means it has the right to hire temporary workers without being obligated to offer them a full-time job after several months.

So rather than blaming the chef for sending you away, tummy grumbling, blame the misguided Socialists.

Eat off the menu and don't be afraid of ordering dishes for which you have to pay a supplement on top of the prix fixe. They're often for very good, local things (fresh cèpes or morels, white asparagus, truffles), or special, labor-intensive dishes of more costly ingredients (at Jacques', lobster fricassée or seared scallops with orange butter sauce, for example), and that's why you're there, hmmm? Ask what is in season. Look at what locals are eating around you. Did they even get a menu?

The French equivalent of a diner around us is the Auberge de la Place in Cazals. Madame Lafon will happily hand you a menu five pages long with six prix fixe menus from seventy-five to two hundred fifty francs, but all the locals are eating Wednesday's *menu du jour* cassoulet at seventy-five francs with wine included. Venture too far off the path (is this really the kind of place in which you want to order tournedos Rossini?), and the further away you get from fresh and the closer you are to the freezer since those recherché dishes aren't ordered very often.

At the Auberge, by the way, for about ten bucks you'll be served a hot, usually brothy, soup, a crudité plate of cold diced beets, grated carrots, and sliced tomatoes or an entrée of the house pâté or terrine, a generous bistro-type main dish with vegetable such as choucroute garnie, the local *mique* or cassoulet, pork chops, steak, and cheese or dessert. Wine and a coffee after are included, too. Your meal will last likely an hour and half in a cheery room filled with local masons and shopkeepers and retirees who exchange lengthy greetings across tables and nod wisely at your choice of the *menu du jour*. The soup arrives in a tureen and the paté in a loaf pan; no one will frown if you take seconds. And your wine carafe will generally be refilled at no charge.

In this region, indeed in my experience throughout much of the unknown southwest, there is an Auberge de la Place in most towns and the larger villages usually marked by a signboard outside with the special of the day. If not, you need only ask where the workers are eating or follow the parade of blue boilersuits come lunchtime.

But know that skipping dishes from a prix fixe does not reduce the cost of the dinner, and that you usually can't substitute this for that or get your main dish with a green salad on the side. The reason it's ten bucks is because, generally, they're making the same thing, with a few variations, for everybody, and there's a lean and mean staff out back that may consist of just two people. This is not to say that certain exceptions aren't possible, however. The one-hundred-fifty-franc prix fixe menu at Jacques' gives you a choice of four or five entrées, main dishes, and desserts. While you can probably order two entrées, the second to be served as a main dish, you can't ask for the reverse. "When a customer asks, 'Would it be possible to have such and such . . . ' " Noëlle says, "rather than assuming it is possible, well, that makes all the difference. We always try to accommodate them, if we can."

And look closely at the menu. Somewhere, usually at the bottom of the page, you will find one of two phrases: *sevice compris* or *prix nets*. The former means that the fifteen percent gratuity is included in the price of every item on the menu, and in the prix fixe meals as well. So you don't have to tip, right, because that fifteen percent is going directly to the waitstaff? Wrong. What actually happens to that money is far less certain. "There are still a few big brasseries, and some of the top, top places where waiters work for a percentage of that," Jacques says. "But that is the exception, and there is nothing in the law any longer about that money going to the help."

Unlike their American counterparts, French restaurateurs are not permitted to pay their staff (except the apprentices) under the standard wage with the idea that the tips will bring the total up to a living wage. They must pay each and every worker at least the *salaire minimum*, currently about $800 a month. When I ventured to Jacques and Noëlle that the tip money was likely going to the staff at most restaurants, they laughed, heartily. The money goes to the owner who passes on what he pleases. In general, after a good meal, you should leave something, even five percent, if you ever want to show your face there again.

Prix nets, on the other hand, means the service charge has not been included, and that the tipping is up to you. Ten percent on top of the bill is considered nice, fifteen percent generous, and leaving it in cash, even if you're paying with a credit card, is very thoughtful indeed. (You will also notice that the credit card print-out, unlike in America, does not have a line for you to enter the gratuity.) It is acceptable to leave nothing after a disaster, and French people do it all the time. Just be sure you're not punishing the waitstaff for the chef or owner's problems. At La Récré, a *prix nets* place, all the hired help—waitress, sous-chef, and dish-washer—share in the tips equally, except Jacques and Noëlle, who take nothing. In the summer, the staff can almost double their salary even before the monthly bonuses that Jacques pays to cover any overtime.

The whole staff, by the way, knows how much you left before the door has closed behind you. And they will, should you turn up unreserved at lunch several days hence, remember and lobby accordingly about whether or not you should be seated. Never have I been more embarrassed as an American than when a group of ten touring bicyclists from the West Coast came through for a meal in the fall and failed to leave a single centime on the table, and this after numerous special requests that the kitchen had honored as best they were able. When all heads turned to me for an explanation after the brightly spandexed crowd had departed, the best I could offer was that we were a young country, still rather coarse and ignorant of more sophisticated European ways.

More modest restaurants will have a limited wine selection and no sommelier, or wine steward. If it is a simple place and a simple meal, you might stick to what comes with it, the local plonk. Or ask the waiter to suggest a better regional wine. La Récréation, for example, has almost every good *vin de Cahors* you could ask for, as well as a number of wines from nearby regions like Gaillac and Buzet. The rest of the list is filled out with a selection of perhaps twenty-five more expensive Bordeaux and Burgundies for those who don't mind spending more for their drink than their food. The whole thing is four pages long.

In a Michelin one-star restaurant, however, the *carte des vins* might be as thick as a phone book and be presented as if it were the bishop's ring. Don't panic. Find the sommelier; he's the guy with the black, usually leather, apron on under his tux and the broken, usually red, capillaries across his nose. (Alcoholism, as well as cavities and gum disease, are the occupational hazards of this work.) He will also be wearing a fetching silver or gold boutonnière representing a bunch of grapes. You might want to ask him to suggest something appropriate, and you're only helping him if you tell him what price level and perhaps what you like or don't like. Ask for *un grand vin de Bordeaux* and watch his eyes light up. Ask for *un vin d'un bon rapport qualité prix*, a good value, and his eyes still might light up as he steers you to one or two of his discoveries.

I had the pleasure one evening, after a very special meal at Le Balandre, a Michelin one-star restaurant in Cahors, of a long conversation with the sommelier, Laurent Marre. When he goes out to a very good restaurant, he always tells the sommelier before they choose the wines how many bottles they're likely to want to drink and what their budget is for them. With this in mind, discussion can then begin about what, exactly, they might prefer.

Ever wonder why the heck they bothered to give you the cork? While the occasional "corked," or spoiled, bottle of wine does happen, especially in older vintages, it is not that common. Smell the cork, which should only ever smell of wine, and you'll know. The cork of a very old bottle, like the 1976 Château Palmer

Margaux a friend opened for us one evening, may even have what looks like a black inkstain around its top, which is a sign of age and no more. The sommelier will give the cork a big sniff, and then he will put it next to your plate before pouring a sip for your approval. Sniff the cork, swirl the wine and smell it, feel its temperature in the glass, hold it up to the light. These are all ways of gathering information before you actually taste it.

If all is to your satisfaction, you do need to give a little nod or word of approval, which is his signal to begin pouring, filling your glass last. If you think the wine is bad, ask the sommelier to try it. Very old red wines do take on a brownish tinge, and sometimes they can become "madeirized." Such wines have an overly sweet bouquet and are thin in the mouth, with no finish, or aftertaste. Sometimes it is simply too old and has lost its essential character; sometimes it is due to poor storage at a too-high temperature. A bad bottle of wine should never appear on your bill. But it is not acceptable to send back a bottle because you don't like it.

If you did choose the grand old Bordeaux, most likely it will be decanted tableside and served in a clear glass carafe. Older wine is decanted both to let it breathe, become oxygenated to bring out its flavors, and because it may have "thrown a deposit," when organic matter, usually harmless crystals of tartaric acid, precipitate out of the wine. Also, the tendency today, especially in America but also among some of the new generation of French winemakers, is to filter less aggressively, leaving more particles to fall out as sediment even in some younger wines.

Diplomacy and curiosity will get you everywhere, too. Is the red wine a little cool? Ask, and you might learn something you didn't know. Some reds, Cahors for example, are drunk slightly cool if they are young, like Beaujolais nouveau. Did your white Châteauneuf-du-Pape arrive warm and yet was opened all the same? Is there a shipwreck of cork debris floating in your wineglass? Did the waiter top off your glass of one red with another before you could stop him? A gentle reminder that this is not how it should be (*Ce n'est pas comme il faut, Monsieur!*) is usually enough if the right thing hasn't already happened. If not, protest, and mightily. You have the right to a fresh bottle and glass.

Sommeliers can be quite forbidding. My wife once asked the very large Robert, who wears the black apron at Le Gindreau, a wonderful if slightly over-the-top Michelin-starred restaurant in the area, if we might have some more bottled water. "Of course, Madame," he responded with a little bow, raising his eyebrows. "By return post!" His little joke at our expense in front of our French friends, while just this side of polite, was a reminder that he was the wine steward, and not the waiter.

Try regional wines, many of which you won't be able to get in the United States—also why you're here, no? This is how you discover that one hundred percent tannat Madiran, the original wine of Gascony, given sufficient age, can be quite a lovely, yet reasonable, mouthful. That dry white Bordeaux of Entre-Deux-Mers from Château le Cluzeau, a château you've never heard of, may be something you'll remember for quite a while.

Meat, staff of life, at least on this side of the Atlantic. The French eat beef *bleu* or *saignant* (nearly raw and rare, respectively) almost always, *à point* (medium) rarely, *bien cuit* (charred) never. Duck breasts, (*cannette*, *caneton*, or *magret*), and lamb and veal are cooked *rosé*, or a little pink, while pork and chicken will be white throughout. The highest-quality meat demands the lightest touch, which is why it is usually seared and crusty on the outside. The crust, in theory, keeps the inside moist and tender because the juices are sealed in. (The best, most fascinating, discussion of these ideas is in Harold McGee's book, *On Food and Cooking*, by the way.) If you don't specify how you want your meat cooked, the chef will cook as he would eat it—generally as little as possible.

What do you do if something goes wrong? If the soup is cold, request that it be reheated, same with the food. Be polite and realize the implications of your request. Namely, if you send your meat back to be cooked more, it is going to take some time and may delay the next course, which will have to be put on hold, for the whole table. Monsieur X, who sends back his main dish without even trying it because the plate isn't hot is asking for trouble, especially if he's already put the kitchen on the spot over the interpretation of "medium-rare." Though cooks and waitstaff are often at loggerheads, when either feels the other's honor is at stake, they stick together. The waitress will not say, "Careful of the plate!" when she returns, and Monsieur X may burn himself on it. His table will likely have to wait for its desserts, as well, his behavior having pushed their priority to somewhere at the end of the line. When a customer puts the staff on the defensive, especially in a loud and very public way, the customer will always suffer. No one is going to spit on your fillet in the kitchen, but there are a hundred little ways in which your good meal will go south very quickly.

At La Récré, Jacques and Noëlle appreciate constructive criticism offered in the right way at the right time. In fact, they often appreciate the judicious suggestion or comment more than the fatuous compliment. I have often seen Noëlle come in halfway through service to remark on some aspect of the cooking, like, "Eh, Jacques, I've seen a lot of people salting the soup and Monsieur X told me he thought it was undersalted, too." While this may not be the nicest thing Jacques has heard all day, he will always go and taste the soup, and, shrugging, add some salt if necessary. Doing dishes, I once commented to Jacques how little leftover food there was on the plates. Most looked like they had been licked clean, with the only debris lobster carapaces and garnish. Almost always, the actual garbage from thirty-five dinners consisted of less than would fit in a one-gallon bucket. When Jacques was trying something new, however, he would sometimes ask the dishwasher at the end of the night what people were leaving on their plates, just to see how it was going over.

People who eat regularly at a restaurant accrue rights that you, if you are a newcomer, do not enjoy. This begins with first names and *bisous* from the staff, a fresh rose on their table, and may carry over to a little plate of off-the-menu *amuse-bouche* little tasties before the soup, and end with a cognac on the house. During the meal, you might notice that they enjoyed an extra slice of lobster tail gracing their lobster-coco salad, were offered a wedge of Camembert instead of the goat cheese on the menu, and that the chef stopped by to share a glass of wine at their table.

Hey, they're paying the same price as I am, why do they get all the special attention? What you can't know is that this is Hervé and Dorothée, a retired couple from Toulouse who've been coming for eight years and that this is their anniversary, for which they've reserved weeks ago. Noëlle remembers that Hervé is allergic to goat cheese and so put aside a wedge of the Camembert from the staff lunch. The rose is because she is fond of them. Jacques knows that Dorothée loves lobster, and the *amuse-bouches* are something he threw in to make the occasion more special.

Although Hervé and Dorothée are fictitious, the elements of the situation as described are not. Noëlle and Jacques (but especially Noëlle) know which of their customers like their soup very hot, which one prefers a bottle of cold Badoit on the table at all times, which one will always take the lobster gratin (so put one aside, Jacques!). Their knowledge, like that of most good, local restaurateurs, extends to their customers' kids, their ages and occupations or where they're at university, recent deaths and births in the family, even sometimes to their last vacation and the kind of car they drive. Noëlle prides herself on recognizing her customers' voices on the phone before they identify themselves. For good customers, they will go some length to cater to their needs, even to the extent of pillaging neighbors' pantries, fridges, and even their gardens in an emergency for bread, cheese, or vegetables if necessary.

The best meals are like conspiracies between chef, waitstaff, and customer. The last Sunday afternoon before the restaurant closed for its winter break in November 2000, a party of eleven, a whole medical practice from Cahors, came up to have their holiday staff meal. For this, Jacques made *coquilles St. Jacques au beurre d'orange*, a gorgeous, very complicated entrée of seared scallops on a bed of mâche with a long-reduced orange juice and butter sauce. It is expensive to make not only because of the scallops, but because it requires a whole crate of fresh oranges and the best, delicate mâche. It is a labor-intensive dish to prepare for a large group because it is complicated to plate and decorate. The mâche has to be artfully laid out, the scallops seared quickly yet more or less at the same time, and then the whole garnished with skinned orange slices, strips of orange peel confit, and petals of confited tomato. And to do eleven at once tends to bring everything else in the kitchen to a full stop.

Why, I had wondered aloud, are you killing yourself on this complicated thing the last day you're open? Jacques shrugged. "They've been coming for a few years now, and I have the time. We're not that busy. It's a lot of work, that dish. But there's one of them, a woman, she liked it so much last time she wrote me a poem. This time," he joked. "I thought maybe she might write a book." And he hardly makes a profit on it.

The best chefs are pigheaded, remember. Pigheaded enough to want to give a customer they've met twice in their life the very real, intense pleasure of a dish the customer loves, even at great inconvenience and with little recompense.

The worst meals are like conspiracies of the chef and the waitstaff against you. Certainly, even very good restaurants have their very bad moments. The chef is sick or on vacation, the head waitress just quit, needed supplies didn't arrive, the walk-in wheezed its last some time late last night, vastly reducing the menu, or your order got lost, overcooked, or otherwise was not what you wanted due to human error. You have every right to complain, politely and at low volume, preferably out of the earshot of other diners. You may want to look at the bill after a major error to see, before you raise your concerns, if they've adjusted it and by how much.

Feel like you want to let them have it? Do you ever want to eat there again? You don't want your name turning up on the *liste rouge*, the blacklist, posted in the back of the reservation book. (Though no restaurant will ever come out and tell a customer he's been banned, that customer will always find the place mysteriously fully booked whenever he calls.)

A few simple things you can do to make everyone happy: Turn off your mobile phone. Bring books, paper and crayons, and quiet games to occupy your kids rather than assuming the waitstaff will occupy them for you.

Social acceptance of smoking in Europe is about where it was in the United States thirty years ago: the worm has not yet turned. The French (and Spanish, and English, and Dutch, and Germans) smoke before the meal, between courses, and after the meal as well. You can ask to be seated near a window or outside, if weather permits, and you can ask other diners if they would mind refraining during your meal. (Good luck.) The French do generally find it reasonable not to smoke next to your baby. If you smoke, others will consider you a paragon of respectability should you ask their permission.

Noëlle, in the cold of fall and spring when she can't open windows, generally leaves ashtrays off the tables. But she has no legal right, even at your behest, to require that party of ten beside you, nine of whom have just lighted up yet again, to extinguish their cigarettes. Otherwise, your only recourse might be a gentle reminder that smoking only dulls the sense of taste. Oddly enough, most people recognize that pipes and cigars belong outside, or at the bar.

If you're traveling to eat (there are many who do!) or prefer your vacation less peopled, consider going out of high season. Spring and fall are beautiful and, with most people back at work, more tranquil, while the museums, attractions, and restaurants are generally still open. At Jacques and Noëlle's (and at many smaller places like theirs), you will find a quieter, gentler atmosphere, and some surprises on the menu. This is because the pace is less hurried, and these are also the seasons when Jacques has the time to try new things, and to offer those dishes that take too much time to prepare for the summer crowds.

One last word of advice. Don't wander into the kitchen unless you're there to fix the dishwasher. Chefs don't like it. Some fellow with a camcorder looming in the doorway, no matter how well intentioned, tends to interrupt the rhythm of the service, not to mention impeding the waitstaff coming and going with hot plates. You might inadvertently ruin someone else's meal. If you want to meet the chef, and he or she does not come out toward the end of service, ask the waitress or barman. Head chefs don't generally do the dessert and cheese course, so they finish a bit earlier and you might catch them at a time when they do not mind spending a few minutes with you. Offer them a drink. Don't overstay your welcome.

GOOD ADDRESSES

Should you want to visit the Lot, and more particularly the area around Les Arques, here is a very brief listing of places to stay, places to eat, and things to do. This part of the country, sparsely populated and undeveloped economically and touristically, offers a more serene, very rural experience more suited to those who prefer long walks to Disneyland. It is also a place where people tend not to speak English, or very little. Unless otherwise noted in the listings below, assume they don't. Widely available guidebooks can tell you about this museum or that architectural wonder (the Michelin Green Guide you want is *Périgord-Quercy*), so instead I've tried to steer a far narrower course to those places to eat and stay, and things to do that you could only learn from someone who's been there.

These listings don't even pretend to be definitive, but they could serve as a starting point for further investigation. More information is available from the Tourist Office of the Canton of Cazals, (tel. 05 65 22 88 88, e-mail otcazals@club-internet.fr), and from various sources mentioned below.

A word on phones. Calling from the United States, dial 011 (international dialing) then 33 (country code of France) plus the number quoted but without the initial zero. When calling in France, also let fall the initial zero. Once you arrive, you may want to buy a phone card at the tobacco shop for use in the phone booths, many of which no longer accept coins.

WHERE TO STAY

Because there is so little tourist infrastructure, there are few hotels. Most lodging is in *gîtes*, separate cottages and houses with kitchens and sometimes a pool. Bed linens and towels may or may not be provided. There are also *chambres d'hôtes*, the French version of a bed and breakfast where the accommodation is attached to but not part of the host house, and they may offer *tables d'hôte*, or meals eaten in a family setting. The nearest luxury hotel is Le Château Mercuès fifteen minutes away, a restored castle perched on a hill above Cahors (tel. 05 6520 00 01, e-mail mercues@relaischâteau.com). There are also small, very rustic village hotels with minimal (very minimal) amenities, all with small restaurants, in Frayssinet-le-Gélat (Le Relais, tel. 05 65 36 66 02), in Cazals (Le Bon Accueil, tel. 05 65 22 81 17), in Pomarède (Le Murat, tel. 05 65 36 66 07), and in Goujounac (L'Hostellerie de Goujounac, tel. 05 65 36 68 67).

What follows is a list of *gîtes* and *chambres d'hôtes*, almost all very local to Les Arques, with English speakers noted at the end of the listing. Some have swimming pools and are quite luxurious done-over old stone houses with all modern conveniences; others are very basic and might seem a bit dowdy to those used to more modern accommodations. (You will rarely find business arrangements with Internet and e-mail in these houses.) As we rented a house for our year, I offer these not as recommendations, but as a starting point, with an emphasis on English-speaking establishments for those linguistically challenged travelers who still seek an authentic, rural experience.

Several organizations which have rental listings located all over the Lot, with more complete information and prices are:

• French Affair, www.frenchaffair.com, an English booking service with a very complete website with pictures of assorted properties and online booking, payment, and reservation from moderately expensive to expensive cottages to villas. Their standards and level of service are very high, they have English area representatives, and the prices reflect this.

- Gîtes de France, ADTRL, address: Maison du Tourisme, Place F. Mitterrand, 46000, Cahors, France, tel. 05 65 53 20 75, e-mail: gites.de.france.lot@wanadoo.fr, 750 listings throughout the department, and you can book through them. English spoken. Catalog available.

- Clévacances-Lot, address: Maison du Tourisme, Place F. Mitterrand, 46000, Cahors, France, tel.: 05 65 53 01 02, e-mail 46@clevacances.com, approx. 300 listings. English spoken. Catalog available at nominal cost.

- Office de Tourisme du Canton de Cazals, address: 46250, Cazals, France, tel. 05 65 22 88 88, e-mail otcazals@club-internet.fr. The Office of Tourism has a listing of local accommodation as well as information on local tourist opportunities. Information from other tourism offices of surrounding cantons in the region: www.quercy.net.

LOCAL GÎTES AND CHAMBRES D'HÔTES

- Domaine des Olmes in Les Arques, two private chambre d'hôtes with small pool, *table d'hôtes* evenings, Peter and Catherine van Buchem, tel. 05 65 21 18 18, e-mail pbuchem@club-internet.fr. English.

- Le Ségalard, Guy and Monique Delaunoit, three private *chambre d'hôtes, table d'hôtes* evenings, small pool, tel. 05 65 21 62 71. Some English.

- Le Coustalou in Marminiac, *chambre d'hôtes, table d'hôtes*, pool, Monsieur Lagarde, tel. 05 65 21 62 32.

- Lesquirol in Les Arques, *gîte* with three bedrooms, Monsieur Pierre Lacombe, tel. 05 65 22 81 52.

- Le Camus, two bedroom stone house in hamlet near Les Arques, reservation through Maison du Tourisme in Cahors, tel. 05 65 53 01 02.

- Profarens, three bedroom stone house about a mile from Les Arques, pool, Monsieur Jean Paul Raymond, tel. 05 65 21 21 71.

- Gagnepo in Cazals, two bedroom stone house w/shared pool outside village, M. Hugues Despriz, tel. 05 65 22 84 77.

- L'Arbre Redon in Goujounac, three bedroom stone house with pool, M. Daniel Soncourt, tel. 05 65 36 67 82.

- Poucaty in Goujounac, *gîte* with two bedrooms, Monsieur Robert Gabarron, tel. 05 65 57 64 78. English.

- Château des Junies ten minutes from Les Arques in Les Junies, two *chambres d'hôtes* in an old château, Mme. Barberet, tel. 05 65 36 29 98. English.

- Stone house with garden in the village of Les Arques, two bedrooms, Monsieur Carrié, tel. 05 65 22 72 89.

- Estanels in Lherm a few miles from Les Arques, five-bedroom stone house with pool, Mr. Brian Hatton, tel. 05 65 21 42 39. English.

- Labergerie de Leobard, in Leobard just up the road from Cazals, contact Gary Heilbronn or Elizabeth Chandler, tel. 05 65 22 85 65, e-mail Labergerie.Leobard@wanadoo.fr, a renovated house (for six) and barn (for eight) with pool. English.

- La Grange at Concores, a little farther afield in the next valley over (fifteen minutes from Cazals), offers four two-bedroom apartments carved out of an enormous, gorgeously converted barn. Reservations: Angie and Richard Atkinson, tel. 05 65 24 59 30, e-mail Angela.Atkinson@wanadoo.fr. English.

• Village house in Cazals, four bedrooms, off main street, tel. 05 65 21 62 59, e-mail TimBranwhite@aol.com.

WHERE TO EAT

La Récréation, in the village of Les Arques, open for lunch and dinner from May 1 through September 30 and weekends (except Sunday dinner) in March, April, October, and November. Reservations: tel. 05 65 22 88 08. Five-course prix fixe menu at about $23 (2001 price) with usually five choices of appetizer, entrée, and dessert plus soup and cheese. Tell them Michael sent you.

Auberge de la Place in Cazals, tel. 05 65 22 86 96, on the central square, is open year-round for lunch and serves dinner in season. A very modest place with various prix fixe menus starting at about $12, it is the place to be in Cazals at lunch. Le Bon Accueil, the hotel just down the street, also has a restaurant open in season with even fewer pretensions.

There are also small restaurants in the surrounding towns of Goujounac (Hostellerie de Goujounac, tel. 05 65 36 68 67), Frayssinet-le-Gélat (Le Relais, tel. 05 65 36 66 02), Pomarède (Murat, tel. 05 65 36 66 07, where the local truckers eat lunch), Salviac (La Casserole, tel. 05 65 41 57 87), and Léobard (Restaurant L'Abbaye, tel. 05 65 41 32 70). None of these restaurants is even in the Michelin Guide, never mind how many forks, and only three, L'Hostellerie de Goujounac, Le Restaurant L'Abbaye, and La Casserole, attempt to offer anything more than a simple sustaining lunch or dinner at a reasonable price.

There are currently five Michelin one-star *restaurants gastronomiques* in the Lot where you can expect to be attended, fed, and cossetted like a king, and at a price, though not so much as you might expect, depending on your choice of wine. Closest to Les Arques is Le Gindreau in St. Médard (tel. 05 65 36 22 27), Le

Balandre (tel. 05 65 30 01 97) in Cahors, and Claude Marco (tel. 05 65 35 30 64) in LaMagdelaine. Le Pont de L'Ouysse (tel. 05 65 37 87 04) and newly starred Le Château de la Treyne (tel. 05 65 27 60 60) are both in LaCave near Rocamadour about an hour north. Le Château de Mercuès, in Mercuès just outside Cahors, recently lost its star but still rates quite highly in the judgment of some.

"Bonnes Tables du Lot," the twice-yearly publication of the Association of Good Tables of the Lot (Alexis Pélissou of Le Gindreau, founder) contains information on more than a score of very good restaurants throughout the département. It is sometimes sold on newsstands and at local chambers of commerce and industry and tourist offices.

A unique experience can be had with a visit to the restaurant of the Lycée Hôtelier Quercy-Périgord, on rue Roger Couderc, tel. 05 65 27 03 00, in Souillac, about an hour to the north of Les Arques. This hotel and restaurant school serves a single four- or five-course prix-fixe dinner menu for about $15. Although the food is fine, you go for the show, for this is where aspiring restaurateurs learn their trade. Your waiter tonight may have been at the stove yesterday, which, as you watch him play drop-the-chop and nearly ignite the table firing up your flambéed dessert, is wholly believable. More seriously, if there were any doubt that classical cooking still prevails, this place erases it when you watch the intensive table-side prep and service of dishes Escoffier would have recognized (and probably invented).

It is also possible to eat farmhouse food at the farmhouse, in so-called *ferme-auberges* and *auberges à la ferme*. These establishments must by law grow and raise on their own land a certain percentage of the produce and meat that turn up on your plate. There are three such establishments local to Les Arques, and they are a good choice for both the most regional of regional cooking and for plain old tying on the feedbag, as the servings tend to be copious. Try Aux Délices de la Serpt in Fraysinnet-le-Gélat (tel. 05 65 36 66 15), La Poule au Pot (tel. 05 65 36 65 48) in Goujounac, or Le Moulin du Roc (tel. 05 65 21 60 51) in Marminiac.

And then, there is always the Michelin. www.michelin-travel.com and available in the United States from your local independent bookseller.

WHAT TO DO

Bring sturdy walking shoes, as the Lot is a walker's paradise. There are few main roads, and tracks through field and forest abound, especially between villages or between hamlets. Many communes have informal walking groups guided by a local. In Cazals, walkers meet every Thursday in front of the Office de Tourisme before hikes from several hours to all day in length. Information and a printed guide of local hikes is available at the Office de Tourisme, 05 65 22 88 88.

Visit the farm. Driving down the one-lane roads, you may come across signs saying *Visite à la ferme*. In and around Les Arques, you can visit foie gras raisers and goat cheese makers and, a little farther afield, many vineyards and even a snail farm. Though they would never say this, I will. Buy something! These busy farmers and growers are taking time to give you a unique look at what they do even though it is not always convenient. Respect their effort by supporting it, even in the most modest way.

Luc and Ghislaine Borie, who figure in this book, live near Catus (tel. 05 65 22 77 81), and their quite extensive and fascinating farm visit and gavage demonstration is at five-thirty P.M. At the Cazals Sunday morning market, there are two producers who sell their products but also welcome you to their Montcléra farms at other times: Philippe Moreau (tel. 05 65 22 87 33) produces foie gras and duck specialties and is the one with the incredible mustaches; there is also Monsieur Pocat-Earl (tel. 05 65 21 69 76) at Farge Bas, who raises organic fruits and vegetables. More local foie gras operations are: La Ferme du Touron (tel. 05 65 36 65 58) in

Goujounac, and Monsieur Perot (tel. 05 65 22 83 86) in Gindou.

At the Sunday market in Cazals, you will see a pleasant, gray-haired woman selling goat cheese and rustic pottery from the same stand. This is Noëlle Maire, and she is happy to welcome you to the farm/workshop La Biquetterie de Terregaye (tel. 05 65 22 70 92) in Thédirac to see the goats and how their milk is made into cheese and to see her own pottery work. Baby goats are born in the early spring, but the animals are personable all year round.

Visit the vineyards. Although *vin de Cahors* is a small appellation, there are more than two hundred AOC vineyards throughout its three growing regions. Unlike the stuffier Bordeaux and California vineyards, these winemakers are generally very happy to show you their entire operation, from a stroll through the vines through the *chais*, the barns where the wine is fermented, filtered, fined, and cask-aged, and finally through the bottling plant and into the caves. If you avoid their busy times, October harvest and July pruning, and always phone ahead for an appointment, you can generally be assured of a long, educational, and winey visit. I have never been charged for a tasting, and only warmly welcomed wherever I have explored, a warmth I have acknowledged with the purchase of a bottle or two at very low prices for very good wine. There are many wine festivals in the summer and fall, dates and locations of which you should be able to find out through the local tourist office or town hall.

One of the best (and least touristed) wine-related events is Le Bon Air Est Dans Les Caves, the Good Air Is in the Wine Cellars Festival in Albas in May (town hall tel. 05 65 20 12 21 for dates). This village climbs a hill on the banks of the Lot, and it was formerly a major wine shipping and storage center. During the festival, many of the wine cellars built into the bottom floors of the houses are open, each one featuring the wine of one vineyard and the music of one group, everything from jazz to a cappella ballads in patois, to gypsy music, to rock 'n' roll. You can subscribe to the modest but filling dinner held down below or pick up foie gras or goat cheese sandwiches as you stroll. The entrance fee is $9 or $10, for which you get a petite tasting glass that serves as your entrée to each cave. Things get progressively wilder as the evening wears on, so you might want to wear your dancing shoes and leave the kids at home.

Without ruling out those special places you may discover or others may recommend, here are ten of my favorite vineyards and wines from all over the Lot:

Clos de Gamot/Château de Cayrou, the Jouffreau-Hermann family, Prayssac, tel. 05 65 22 40 26. English.

Clos la Coutale, Philippe Burnède, Vire-sur-Lot, tel. 05 65 36 51 47. English.

Clos Triguedina, Jean-Luc Baldès, Puy L'Évêque, tel. 05 65 21 30 81, English.

Château de Chambert, Marc and Joel Delgoulet, Floressas, tel. 05 65 31 95 75.

Château Paillas, Germain Lescombes, Floressas, tel. 05 65 36 58 28.

Château Pineraie, Jean-Luc Burc, Puy l'Évêque, tel. 05 65 30 82 07.

Château du Cèdre, Verhaeghe and son, Vire-sur-Lot, tel. 05 65 36 53 87.

Château de Haute-Serre, Cahors, Vigouroux family, tel. 05 65 20 80 80.

Château Saint-Didier Parnac, Franck and Jacques Rigal, Parnac, tel. 05 65 30 70 10.

Christian Belmon in Goujounac, tel. 05 65 36 68 58, also makes an out-of-appellation *vin de pays* that is unusual and delicious, if relatively pricey at about $10 a bottle.

(If you're coming from, or heading to, Bordeaux, by the way, after you've tasted at one or two of the big dogs, consider a visit to Thierry Deschamps and his wife, Luze, at their Château Le Cluzeau, tel. 05 57 84 01 06, in Mérignas, Gironde, a small

Bordeaux Superieur producer who really puts out the red carpet.)

And if you can't make it to a vineyard but still want to find a very good selection of local wine and particularly older bottles which are ready for drinking, go to La Maison des Vins du Valentré (tel. 05 65 22 12 22) in Cahors, just at the foot of the famed Valentré Bridge. Closer to Les Arques, Claude and Lydia Bouysset have a reasonably varied selection at the Épicerie Bouysset in Cazals, and there is Les Vins de la Vielle Treille (tel. 05 65 41 57 57) in Salviac.

Go riding. One of the most incredible ways to explore this landscape is on horseback. Geneviève Crassat and Pierre-Jérôme Atger run La Fontaine in Goujounac (tel. 05 65 36 69 63 or 05 65 36 63 94, e-mail: la.fontaine@libertysurf.fr). This young couple, who both speak some English, offer relaxed riding experiences from a morning's hack to six-day outings through Quercy and Périgord organized around various themes, including a gastro-nomic and wine tour through the Cahors vineyard country. They provide everything, and many of their horses are gentle, sturdy Pyrénéean Mérens breed. There is also the Equestrian Recreation Center in Frayssinet-le-Gélat, tel. 05 65 36 61 46, which offers a more formal riding experience and lessons. Riding in France, at about one-third of the cost in the United States, is one of the best bargains around.

Go to the markets. In every canton there is generally one mar-ket, usually in the largest town. They are staggered throughout the week for your convenience. They usually open at seven-thirty or eight A.M. and are gone by noon or one P.M. although there are evening markets, too. Catus has its morning market on Wed-nesdays, but has some evening hours, five P.M. to midnight, in July and August. Sunday, the market is in Cazals, Wednesdays in Salviac, with organic products featured there every second Sunday as well. The widest variety, and not only of fruits, vegetables, fish, meat, charcuterie, and duck products, is found in the biggest mar-kets, so venture down to Puy l'Évêque or Prayssac. Market days and special events in other towns can be found at www.quercy.net. A real treat—and a place to find unusual ingredients—is the Les

Halles in Cahors (the big glass-domed building on rue Clémenceau off Boulevard Gambetta), open six days a week (Monday closed) and bursting with all good things. Be sure to visit Monsieur Marty's creamery, where he'll not just ask you which of his incredible cheeses you want, but when you intend to serve it so he can make sure it's perfectly ripe.

There are markets and fairs for everything, not just your weekly groceries. These events come and go with the seasons. For example, there are truffle markets in Martel and in Lalbenque, which, although they sell only to wholesalers, are a great spectacle. Because truffles are a winter phenomenon, these markets are only open from November to February. In the summer, there are market festivals in the honor of the melon, distilled spirits, foie gras and duck products, livestock, poultry, antiques and junk, plants and flowers, books old and new, almost anything someone may want to buy.

Go to the fêtes. Every flyspeck of a village has its summer fête, generally three days of events starting on Friday and culminating in a communal meal and dancing with live music and fireworks on a Saturday or Sunday night. Pick up a copy of Sunday's *La Dépêche*, the regional newspaper (also online at www.ladepeche.com). In the "Lot" section, the newspaper provides, commune by commune, a listing of what is happening that week with contact information. La Fête des Arques is usually held the second weekend in August, and you can subscribe for the simple meal at the town hall, tel. 05 65 22 87 55.

Cazals, Catus, and Frayssinet-le-Gélat all have *plans d'eau*— artificial lakes with beaches, snack bars, playgrounds, sometimes basketball and tennis courts, acres of grass and trees, and, especially in the summer, lots and lots of children to play with your kids. They are a place to go when all other possibilities run out, and you'd be surprised how quickly your child makes friends and begins to speak a bit of French. Language doesn't have to be a barrier, either, as there are many English visitors, and the Dutch and Germans tend to get along, too. Sipping a cool drink under an umbrella while your kids romp in the water under the eyes of the lifeguard, you may meet some interesting people.

Above all, walk. The motto of the Lot is "A surprise at every step," and happily the vast majority are pleasant ones. So lose yourself in the village byways and farm roads, in the forests and fields. Take that inviting track through the chestnuts. Don't be surprised if the locals are interested in you and want to talk. These are very friendly, curious people who will return twofold any effort you make.

Comments and suggestions? Write the author at theyard@ gwi.net.

REFERENCES

Auricoste, Françoise. *Les Arques en Quercy: Vallée du fer, Vallée des arts. (Les Arques in Quercy: Valley of Iron, Valley of Arts.)* Cazals, France: Syndicat d'Initiative de Cazals, 1990.

Béghain, Patrice. *Écrivains et artistes en Quercy. (Writers and Artists in Quercy.)* Rodez, France: Éditions du Rouergue, 1999.

Béziat, Marc. *Recettes Paysannes du Lot. (Country Recipes of the Lot.)* Rodez, France: Les Editions Du Curieux, Rodez, France, 2000.

Bové, José and François Dufour. *Le Monde n'est pas une Marchandise. (The World Is Not a Commodity.)* Paris: Éditions la Découverte, 2000.

Christine Bonneton Éditeur. *Encyclopédie Bonneton du Lot.* Paris: Christine Bonneton Éditeur, 2000.

David, Elizabeth. *Elizabeth David Classics.* Newton, Mass.: Biscuit Books, 1998.

———. *French Provincial Cooking.* London: Penguin, 1972.

Dulvat, Claudine and Jeanine Pouget. *Recettes du Quercy. (Recipes from Quercy.)* France: Les Editions du Laquet, 1991.

Dyson, Henry. *French Real Property and Succession Law.* London: Robert Hale, 1988.

Jost, Philippe. *La Gourmandise: Les chefs-d'oeuvre de la littérature gastronomique de l'Antiquité à nos jours. (La Gourmandise: Masterpieces of Gastronomic Literature From Antiquity to Our Time.)* France: Éditions Le Pré aux Clercs, 1998.

Jouffreau, Jean. *La Passion faite Vin . . . de Cahors. (A Passion Called Wine . . . the Wine of Cahors.)* Cahors: France: privately published, 1993.

Kelly, Amy. *Eleanor and the Four Kings.* Cambridge, Mass.: Harvard University Press, 1978.

Lang, Jenifer Harvey, ed. *Larousse Gastronomique.* New York: Crown, 1988.

Larousse Staff, eds. *Larousse Wines and Vinyards of France,* London: Ebury Press, Random Century Group, 1990.

L'Hôtellerie Magazine, October 5, 2000, pp 4–7 and 14–18.

Marks, Claude. *Pilgrims, Heretics, and Lovers: A Medieval Journey.* New York: Macmillan, 1975.

McGee, Harold. *On Food and Cooking: The Science and Lore of the Kitchen.* New York: Fireside/Simon & Schuster, 1984.

McPhee, John. "Brigade de Cuisine" in *Giving Good Weight.* New York: Farrar, Straus, Giroux, 1983.

Michelin, eds. *Guide Michelin Rouge 2001.* Paris: Éditions des Voyages, 2001.

Olney, Richard. *Simple French Food.* New York: Macmillan, 1974.

Prax, Valentine. *Avec Zadkine: Souvenirs de Notre Vie. (With Zadkine: Memories of Our Life.)* Paris: La Bibliothèque des Arts, 1995.

Root, Waverly. *The Food of France.* New York: Vintage Books, 1992.

Strang, Jeanne. *Goosefat & Garlic: Country Recipes from South-West France.* London: Trafalgar Square, 1994.

Strang, Paul. *Wines of South-West France.* London: Trafalgar Square, 1995.

Sumption, Jonathan. *The Hundred Years War: Trial by Battle.* London: Faber and Faber, 1999.

————. *The Hundred Years War: Trial by Fire.* Philadelphia: University of Pennsylvania Press, 1999.

White, Freda. *Three Rivers of France: Dordogne, Lot, and Tarn.* London: Faber and Faber, 1972.

Wolfert, Paula. *The Cooking of South-West France.* New York: Harper & Row, 1983.

Zadkine, Ossip. *Le Maillet et le Ciseau: Souvenirs de Ma Vie. (The Mallet and the Chisel: Memories of My Life.)* Paris, France: Éditions Albin Michel, 1968.

RECOMMENDED READING

An annotated and expanded version of this list, as well as a list of cookbooks, menus, and sources for ingredients relating to southwestern French cuisine, can be found on the author's web page: www.gwi.net/~theyard.

Bemelmans, Ludwig. *La Bonne Table.* New York: Simon & Schuster, 1964.

Brillat-Savarin, Jean-Anthelme. *The Physiology of Taste.* London: Penguin, 1970.

Cooper, Artemis. *Writing at the Kitchen Table: The Authorized Biography of Elizabeth David.* New York: Ecco Press, 2000.

Daley, Robert. *Portraits of France.* Boston: Back Bay Books/ Little Brown, 1991.

Echikson, William. *Burgundy Stars: A Year in the Life of a Great French Restaurant.* Boston: Little, Brown, 1995.

Fisher, M. F. K. *The Art of Eating.* New York: Macmillan, 1990.

Freeling, Nicholas. *Kitchen Book: The Cookbook.* Boston: David R. Godine, 1991.

Goodman, Richard. *French Dirt: The Story of a Garden in the South of France.* Chapel Hill, N.C.: Algonquin Books/ Workman, 1991.

Gopnik, Adam. *From Paris to the Moon.* New York: Random House, 2000.

Graham, Peter. *Mourjou: The Life and Food of an Auvergne Village.* New York: Viking/Putnam, 1998.

Hesser, Amanda. *The Cook and the Gardener.* New York: W. W. Norton, 2000.

Kurlansky, Mark. *Cod: A Biography of the Fish That Changed the World.* New York: Walker, 1997.

Liebling, A. J. *Between Meals.* New York: North Point Press, 1986.

Mitchell, Joseph. *Up in the Old Hotel.* New York: Random House, 1992.

Oyler, Philip. *The Generous Earth.* London: Hodder & Stoughton, 1950.

———. *Sons of the Generous Earth.* London: Hodder & Stoughton, 1963.

Reichl, Ruth. *Tender at the Bone.* New York: Broadway Books, 1999.

———. *Comfort Me with Apples.* New York: Random House, 2001.

Rosenblum, Mort. *A Goose in Toulouse and Other Culinary Adventures in France.* New York: Hyperion, 2000.

Ruhlman, Michael. *The Soul of a Chef.* New York: Viking Press, 2000.

Sokolov, Raymond. *Fading Feast: A Compendium of Disappearing American Regional Foods.* Boston: David R. Godine, 1992.

Steingarten, Jeffrey. *The Man Who Ate Everything.* New York: Vintage, 1998.

Stevenson, Robert Louis. *An Inland Voyage, Travels with a Donkey, and The Silverado Squatters.* New York: Dutton, 1925.

Tendret, Lucien. *La Table au Pays de Brillat-Savarin. (Eating in the Land of Brillat-Savarin.)* Lyon, France: Éditions Horvath, 1986.

Thorne, John. *Serious Pig: An American Cook in Search of His Roots.* New York: North Point Press, 1996.

———. *Outlaw Cook.* New York: North Point Press, 1994.

Wechsberg, Joseph. *Blue Trout and Black Truffles.* Chicago: Academy Chicago Publishers, 2000.

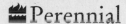 Perennial

Books by Michael S. Sanders:

FROM HERE, YOU CAN'T SEE PARIS
Seasons of a French Village and Its Restaurant
ISBN 0-06-095920-7 (paperback)

A sweet, leisurely exploration of the life of Les Arques (population 159), a hilltop village in a remote corner of France untouched by the modern era. It is a story of a dying village's struggle to survive, of a dead artist whose legacy began its rebirth, and of chef Jacques Ratier and his wife, Noëlle, whose bustling restaurant—the village's sole business—has helped ensure the village's future.

"[Sanders] renders the restaurant's workday as cannily as he does the village's moments of abrupt dislocation from the present, when the air suddenly seems to hold a thousand years of history in it." —*Kirkus Reviews*

THE YARD
Building a Destroyer at the Bath Iron Works
ISBN 0-06-092963-4 (paperback)

Illustrated throughout with dramatic photographs and detailed drawings, *The Yard* is not only a fascinating inside look at one of the greatest shipyards in the United States, but also a very human chronicle of the lives of those who work there, and a tribute to the end of an era.

"[Sanders does] an admirable job of conveying the life history of a destroyer. . . . He is a superb expository writer." —*New York Times Book Review*